The TINA Book

The TINA Book

A co-operative solution for a competitive world

Hendrik Berndt
Emmanuel Darmois
Fabrice Dupuy
Motoo Hoshi
Yuji Inoue
Martine Lapierre
Roberto Minerva
Roberto Minetti
Cesare Mossotto
Harm Mulder
Narayanan Natarajan
Max Sevcik

and

Martin Yates

Prentice Hall Europe

London New York Toronto Sydney Tokyo
Singapore Madrid Mexico City Munich Paris

First published 1999 by
Prentice Hall Europe
Campus 400, Maylands Avenue
Hemel Hempstead
Hertfordshire, HP2 7EZ
A division of
Simon & Schuster International Group

Typeset in 10/12pt Times
by Fakenham Photosetting Limited

Printed and bound in Great Britain
by Redwood Books, Trowbridge, Wiltshire

Library of Congress Cataloging-in-Publication Data

Available from the publisher

British Library Cataloguing in Publication Data

A catalogue record for this book is available from
the British Library

ISBN 0-13-095400-4

---

1 2 3 4 5 03 02 01 00 99

Contents

Acknowledgments

This book would not have been possible without the work of a very large number of people. We express gratitude to all Associations we have partnered and whose members have given most valuable feedback on the TINA architecture: European Commission Research Program ACTS, EURESCOM, OMG, ITU-T, NMF, DAVIC and the ATM-Forum.

Among the people most responsible for establishing the TINA Consortium, we express our indebtedness to Tom Rowbotham (BT); Sadahiko Kano (NTT); Gary Handler and Deb Guha (Bellcore). We thank all members of CSB and CTC, the legal expert group headed by Martin Read (BT), and Mario Campolargo (EC) for his co-operation from the onset. In addition we wish to thank the following persons for a thorough review of parts of the book: Guiseppe Giandonato, Michel Ruffin, Patrick Hellemans, Patrick Farey, Telma Mota, Philippe Terrien, Corrado Moiso, Carlo Licciardi and Joël Seguret. We also wish to thank Nicolas Mercouroff, Anastasius Gavras and Helen Tourelle for their help in the redaction of parts of the book.

From 1993 to 1997, many engineers from the member companies have been working in the Core Team in New Jersey. They are:

Abarca, Chelo (Alcatel SESA), Addagarla, P. Satyaswarup (Bellcore), Ashford, Colin (Northern Telecom), Aurrecoechea, Cristina (Columbia University NYC), Bagley, Mark (BT), Berndt, Hendrik (Deutsche Telekom AG), Bloem, Jack (Unisource (former KPN Research)), Brown, Dave (DSC Communications (former NEC)), Chapman, George (Bell Sygma Telecom Solutions), Chapman, Martin (IONA (former BT)), Choi, Eunho (Korea Telecom), Christensen, Heine (Tele Danmark), Cohen, Viviane (Lucent (former France Telecom, OCTEL)), Colban, Erik (Ericsson AS (former Telenor)), Darmois, Emmanuel (Alcatel), Dathi, Naiem (Hewlett Packard), DeMaio, Joseph A. (Bellcore), Demounem,

Laurence (Alcatel), Dupuy, Fabrice (France Telecom), Farley, Patrick (BT), Feldman, Stuart (IBM (former Bellcore)), Flinchbaugh, Glenn (The Anderson School at UCLA (former Bellcore)), Flinck, Hannu (Nokia), Forslow, Jan (Ericsson), Fratini, Stephen (Bellcore), de la Fuente, Luis Alberto (Telefonica I+D), Garcia Lopez, Juan Carlos (Telefonica I+D), Garcia Lopez, Eduardo (Price Waterhouse (former Telefonica I+D)), Gatti, Nicola (CIE Cegetel (former Telecom Italia)), Gavras, Anastasius (Deutsche Telekom AG, Technologiezentrum), Goutet, Matthieu (HP), Graubmann, Peter (Siemens), Gutierrez, Raul (Telefonica I+D), Hall, Milton (Bellcore), Hamada, Takeo (Fujitsu), Hamdan, Abdul Hamid (Telekom Malaysia), Hammainen, Heikki (Nokia Telecommunications), Handegard, Tom (Telenor), Hansen, Per Fly (Tele Danmark), Hara, Hiroaki (NEC), Hedberg, Magnus (Telia), Hegeman, Johannes (KPN Research), Hogg, Stephanie (Telstra), Hopson, Alan (Telstra), Horrer, Matthias (Alcatel), Hoshiai, Takashige (NTT), Hoshi, Motoo (NTT Advanced Technology (former NTT)), Hwang, Woo Sun (Korea Telecom), Ikegawa, Takashi (NTT), Jansson, Rickard (Ericsson), Jeon, Hong Beom (Korea Telecom), Jorgensen, Mikael (CRI A/S (former Tele Danmark)), Kamata, Hiroshi (OKI), Kanasugi, Keiji (Fujitsu), Kanishima, Ken (NTT), Kantola, Raimo (Helsinki University of Technology (former Nokia)), Kavak, Nail (Telia), Kawanishi, Motoharu (OKI), Kawanobe, Akihisa (NTT), Kelly, Edward (First Union National Bank (former DEC)), Kim, Cheol Koo (Samsung), Kim, Dae Seok (Samsung Electronics), Kim, Hyun Cheol (ETRI), Kim, Sangkyung (Korea Telecom), Kitson, Barry (Telstra), Kiwata, Kazuhiro (NTT), Kobayashi, Hidetsugu (NTT), Kristiansen, Lill (Telenor), Kudela, Melanie (Lucent Technologies (former Stratus/Isis Distributed Systems, AT&T)), Lee, Dongmyun (Korea Telecom), Lee, Han Young (Korea Telecom), Lengdell, Magnus (Ericsson Telecom (former Telia)), Leijdekkers, Peter (KPN Research), Licciardi, Carlo Alberto (CSELT), MacKinnon, Ken (Northern Telecom), Mampaey, Marcel (Alcatel), McMordie, W. Shane (BT), Mercouroff, Nicolas (Alcatel), Minerva, Roberto (CSELT), Minetti, Roberto (CSELT), Montesi, Stefano (Telecom Italia), Montibello, Alphonso (IBM), Moore, Kathryn (BT), Moreno, Juan Carlos (Telefonica I+D), Mulder, Harm (PTT Telecom B.V.), Nagasaka, Mitsuru (Hitachi), Nakashiro, Koki (Hitachi), Natarajan, Narayanan (Bellcore), Nilsson, Gunnar (Ericsson), Nordin, Mats (Telia), O'Neil, Joseph (AT&T), Ohtsu, Kazuyuki (Hitachi), Oshigiri, Hiroshi (NEC), Parhar, Ajeet (Telstra), Pavon, Juan (Alcatel), Pensivy, Stephane (France Telecom), Plummer, Ron (TINA-C), Prozeller, Paul E. (AT&T), Rajahalme, Jarno (Nokia), Richter, Lars (Telia), Rosli, Raja Mohd (Telekom Malaysia), Ruano, Fernando (Telefonica I+D), Rubin, David (Lucent), Rubin, Harvey (Lucent Technologies (former Bellcore)), Sallros, Johan (Telia), Schenk, Mike (KPN Research), Singer, Nikolaus (Alcatel), Speirs-Bridge, Andrew (Telstra), Spinelli, Graziella (CSELT), Steegmans, Frank (Alcatel), Takita, Wataru (NTT), Tani, Hideaki (NEC), de Tournemire, Eric (France Telecom), Utsunomiya, Eiji (KDD), Wakano, Masaki (NTT), Walles, Tony (BT (retired)), Watanabe, Nobuyuki (NTT), Westerga, Richard Simon (KPN), Wheeler, Geoff (Telstra), Yagi, Hikaru (KDD), Yang, Henry (Xiong) (Lucent (former IBM)), Yates, Martin (BT), Yun, Dong Sik (Korea Telecom).

There have been and still are many engineers working within the member companies on Auxiliary Projects, World Wide Demo and TTT activities. We shall not name anyone here, but express our deepest gratitude to all of them.

Introduction

The Telecommunications Information Networking Architecture (TINA) is now an 8-year-old adventure, with an official debut in a TINA Workshop in 1990. It started quite informally as a network of telecommunications specialists (in particular, but not only, from the area of intelligent networks) that wished to co-ordinate among their companies a variety of scattered, advanced research. The development of TINA had been dramatically accelerated by the creation of the TINA Consortium (TINA-C) by the end of 1992. TINA-C was created with a main aim of producing a set of architectural software specifications validated by various experiments. This work relied greatly on the TINA Core Team (CT), a permanent group of engineers; though they were transient, insofar as they were regularly replaced by other colleagues coming from most of the member companies, their achievement has been impressive. The TINA Consortium also relied on parallel projects, the so-called Auxiliary Projects; these were undertaken in the member companies: "at home" so to speak. This community now consists of several hundred adepts; we can mention only the more than two hundred engineers that have been participating in the Core Team, going beyond the limits of the TINA Consortium itself, as a network of "friends and family" of TINA among competitors and customers.

After more than five years the TINA Consortium has now changed to a TINA Forum: engineers meet in working groups where they exchange the result of the "at home" research and development, and they prepare pre-standardization documents and strategy to get the TINA concept standardized by various bodies. The TINA Forum includes a TINA Architecture Board that ensures coherence of the various specifications.

TINA-C has generated many results. Among them, a number of "official" TINA-C documents that reflect the consensus of the member companies after thorough reviews based on the validation experience acquired in the Core Team and the Auxiliary Projects:

those "Baselines" are the core of TINA, its valuable treasure. In addition, TINA-C is also an enormous effort of conceptualization, specification, criticism, validation, construction and deconstruction. In short, the activity of a large "think tank" of hundreds of people. Needless to say, TINA is also composed of this more "informal" set of results, this accumulation of various points of view and list of open issues.

The aim of this book is to introduce to this various, wide (and sometimes complex) set of concepts, architectures and experiments that reflect future visions of the information technologies and telecommunications industries. It is based on the most recent results of TINA-C but is not meant to be an "official" presentation of TINA-C. All the authors are now (or have been recently) playing an important part in TINA-C. However, they wanted to bring to this book a broad perspective on TINA and the future of information networking.

WHAT IS IN THE BOOK

Not surprisingly, the book takes a top-down approach (it will appear in the remainder of the book that this is the TINA state of mind) from explaining the requirements for a new software architecture for information networking to the gradual presentation of TINA concepts and how they will be applied.

Chapter 1 introduces the requirements posed by the emerging industry of information networking services. Themes such as multimedia, the Internet and evolution needs of the traditional telecommunications, for example intelligent networks and personal mobility, are analyzed.

Chapter 2 is a brief presentation of TINA. It is a survey, presenting the main concepts, which is expanded in the next four chapters. Some of the modeling concepts used by TINA are introduced more precisely (they are not elaborated further in the book) and the main characteristics and components of the three sub-architectures are presented. Pointers to the vast TINA literature are given.

Chapter 3 is a presentation of the computing architecture. It first clarifies the object model and introduces the TINA Object Description Language (ODL) in comparison with OMG's IDL. The distributed processing environment (DPE), a series of concepts that presents a uniform view of heterogeneous computing resources, is then discussed. The nature of the DPE is discussed in comparison with OMG's CORBA, a strong candidate technology on which the DPE could be built.

Chapter 4 describes the service architecture which offers a set of concepts, objects and mechanisms for services to be built upon. A very important aspect of service architecture is the notion of session that is described in detail. In many respects, this is one of the most important and original facets of TINA. This presentation is based on an example.

Chapter 5 describes the network resource architecture (NRA) and explains the notion of "TINA networks". The focus is in part on the description of how network resources may be treated to present a common scheme for control and management. The NRA describes the way requirements imposed on TINA for multimedia services are met. Here again, an example allows a comprehensive explanation of the various concepts introduced.

Chapter 6 explains the TINA model used to describe conformance of a system to TINA. The main concepts are the business model and the reference points. The main reference points, those currently subject to the specification process in TINA-C, are presented in more detail.

Chapter 7 analyzes different approaches to the way TINA can be introduced into networks. In the previous chapters TINA has been presented in isolation as a "pure" solution which, in reality, will never be the case. This chapter analyzes the types of interworking or migration strategy that are required. There are various target networks for the introduction of TINA and there is no unique approach, even for a given network (e.g. intelligent networks or the Internet), so different scenarios are presented.

Chapter 8 addresses implementation of TINA. It analyzes how a methodology might be applied in the various phases of the service development lifecycle. Some of the current efforts to make TINA practical are presented regarding the development of techniques and tools to support TINA concepts. An insight into the large-scale demonstration of TINA planned in 1998 (The TINA Trial) is also offered.

Chapter 9 concludes with an assessment of TINA's major strengths. This reflects TINA's maturity and also underlines a work program for a constantly evolving TINA in the next four years.

Appendix 3 contains a brief introduction to the main background required for this book (object orientation, CORBA, intelligent networks, TMN).

WHAT TO READ

There are different ways to read this book. Those who want an "executive" view of TINA can read Chapter 2 (TINA in a nutshell) and Chapters 1, 7 and 8 for understanding which requirements TINA is made to answer, and how TINA will be deployed. For a more in-depth approach, Chapters 3 to 6 are more detailed and make a good technical introduction to TINA. There is more to read in the large amount of TINA-C literature.

The authors have attempted to present the concepts in a way that is understandable to readers who have an interest in IT or telecommunications.

Chapter

1 New telecommunications business opportunities

Telecommunications Information Networking Architecture stands at the crossroads of two industries whose convergence has long been announced but not yet achieved: the telecommunications industry and the information technologies industry. It is this trend, among others, that the originators of TINA-C had in mind when they created the TINA Consortium and put together the so-called Core Team that worked on TINA from 1993 to 1997.

The basic assumption behind TINA is that no solutions to the challenge posed by the emergence of the new information society will be found without putting together the best of telecommunications and information technologies. TINA is at first an attempt to propose a synthesis between the two worlds: the purpose of this book is to present readers with enough elements to convince them that this synthesis is under way, at least regarding TINA's major objective. This is to provide a coherent and common software architecture for the networked information services that will yield most of the revenue from telecommunications at the dawn of the twenty-first century.

This high-level description of the scope of TINA is sufficient to make it clear that, though this architecture is broad, it is meant to answer a certain number of requirements but not all of them at the same time. TINA is not meant to be and is not going to be a "big bang" in telecommunications. This chapter will analyze some of the main trends in telecommunications and underline the approach taken by TINA from the very start to address them.

Two main aims of TINA-C have been to find the means to speed up the evolution of existing "legacy" (though sometimes very recently deployed) networks, and to help investigate support for the creation of new revenues through new services. They have guided the technical approach followed by the various participants. But probably the most important goal of TINA-C was to create the foundations for collaboration between the different players to share some of the risks of new business creation. The global situation of telecommunications has evolved dramatically in the last five years but this even more justifies TINA-C advocating "a co-operative solution for a competitive world".

From the analysis of the major trends in telecommunications, this chapter will underline the requirements that are placed on TINA. A first assessment of the main differences between TINA and other architectures will be proposed. The following chapters will show how these requirements, whether technical or business, have been (or have not yet been) met.

1.1 TRENDS AND OPPORTUNITIES

Telecommunications have a history of more than a century. During this period, they have continued to provide a communications service simple enough to be usable by anyone and are firmly established as an indispensable infrastructure. Today they provide services for more than 800 million terminals around the world and enable connections to locations in any country at any time by a simple process of dialing. The terminals connected have grown not only in number but also in variety: telephones, facsimiles, computers, etc. Inside the telecommunications network, a vast number of telecommunications systems scattered around the world operate under the control of software. The current telecommunications network is the largest on-line, real-time distributed processing system ever created. Large and complex as it is, worldwide interconnection has been made possible by the co-operation of many network operators and vendors in setting and observing international standards.

1.1.1 Evolution of services

Recently the arena of telecommunications has broadened as users are freed from wires by the use of wireless telephone and satellite communications. Mobility will become increasingly a standard feature that users will want in any network. This will affect the nature of networks. Commonality of solutions will be a significant requirement.

Corporate/enterprise networks have become critical to businesses as information has become a vital strategic weapon. They employ high-speed leased lines and require global and seamless networks. To meet such new developments, telecommunications providers see it as their natural mission to provide a one-stop shopping service for customers to get end-to-end connections across national boundaries. They deploy intelligent networks (IN) and other mechanisms within their networks to prompt the provision of new services and enable the management of the vast network resources. ISDN, ATM and other high-speed access technologies have been employed and serve as the platform for supporting the rapid expansion of the Internet.

1.1.2 Changes in the business environment

Apart from these technical developments, the telecommunications industry was ushered into a new business environment about a decade ago. Bidding farewell to a long period of national enterprise or natural monopoly, the telecommunications industry faced deregulation and introduction of competition. Together with the need for global, seamless networks, the competition has mobilized dynamic relationships between network providers. These trends are not temporary but are expected to continue for some time.

Deregulation will have many consequences for organization of business in the next decade. One of the most significant changes will be induced by the separation of service delivery from network provision. This will significantly modify the nature of the public network operators' business.

The relationship between network providers and vendors has also experienced changes. The political maneuvring of the telecommunications industry of a decade ago is now replaced by the principle of a competitive market. The currently accepted concepts are multicarrier and multivendors.

In addition, IT vendors have become involved in the telecommunications industry. A new class of providers, called Internet service providers (ISPs), has arisen and is growing rapidly. A significant characteristic of services in this business area is that they are not meant to meet the usual telecommunications requirements regarding quality of service (QoS). The QoS of these services is dependent on the actual traffic and will only give best-effort performance.

A very good example of business change and corresponding opportunities is that of the USA. Here the environment has been changed considerably by the Telecommunications Act 1996. One very substantial consequence of this law is the opening of the local telecommunications market to competitors, including long-distance carriers and, vice versa, allowing local companies to enter the long-distance business under terms that prove that competition in their local area is granted. Until then, the existing seven Regional Bell Operating Companies (RBOCs) in the USA found themselves in a monopoly-like situation, whereas the long-distance business has been exposed to competition for years.

This situation is visible from a pricing perspective: the vast majority of RBOCs' customers used to be charged flat rates for local services. Price structures for the local loops have so far to be understood in terms of universal coverage, which means that "real" costs are somehow subsidized. Therefore it became a major undertaking to outlay prices for local loops in a competitive environment. It is not an exaggeration to state that this process of opening the local market for competition turned out to be an unexpected hurdle race in which the CLECs (competitive local exchange carriers) have to compete against the ILECs (incumbent local exchange carriers). Recent forecasts on revenue figures of some of the long-distance carriers in the USA seem to prove this observation by being lower than expected because of unappreciated gains in local business.

But there are more business opportunities in the local area: as the local loop is utilized to grant equal access to any service provided in either the telco or the Internet domains, new opportunities occur. Furthermore, based on new network technologies for high-speed local access (like the Asynchrony Digital Subscriber Loop (ADSL) with 1.5 to 6 Mbit/s downstream and 0.5 to 1.5 Mbit/s upstream), new interactive services can be "shipped" to customers' homes. They create new virtual homework environments based on broadband services and more. There is a chance for newcomers to the local US market not approaching it with an "me-too" attitude but willing to distinguish themselves from the ILECs.

Many other legal issues are pending that are considered to drive the US market, such as those related to electronic commerce, Internet tax regulations, encryption and privacy, electronic mailbox protection and unsolicited commercial electronic mail choice, to name a few.

1.1.3 The revolution of information networks: towards a multimedia era

Conventionally, the basic telecommunications service has been to provide a transport means called a "circuit". Recently, services have become more varied, including World Wide Web (WWW) and electronic mails that now have become basic services as well. Accordingly, the network can now be termed an information network.

This new information network has been built to foster a "cyberspace", a virtual society, where various leading-edge experiments are made to examine the viability of new information network services. Telemedicine, telelearning, electronic commerce employing electronic money and virtual shops are rapidly becoming viable propositions.

If the telecommunications network is to continue to meet social demand, it should grow not only on its traditional track but also off the beaten path. Three major shifts are taking place that are dramatically changing the culture of the telecommunications industry.

The first is a change in the rationale for offering services. A traditional approach has been to ask what services can be provided with the existing network and add the required features to the network in a well-planned manner. The new approach is to ask what services are wanted and to find required solutions in a timely fashion. There is a growing need for a service provider to be able to try as many services as possible, then to assess the success of each of them. The winning ones will be added to the customer's portfolio: new services will start small at low cost and the coverage will be extended as soon as the service becomes successful.

The second is a change in the criteria for a good network. The traditional criteria have revolved around building an economical network. The new criteria are concerned with how to build a network rapidly that can provide diverse services and is open for interconnection and future development. The shift has become sensible because of the great reductions in long-haul transmission costs and the availability of efficient communications technologies, such as ATM.

The third is a shift in the role played by the network. The conventional network has primarily provided connection services. But the network should now acquire the capability for networking between heterogeneous systems and operators, to provide dynamic services. The network becomes only one element in the overall platform for providing application-oriented services.

1.2 MORE ABOUT THE CHALLENGE OF NEW SERVICES

1.2.1 The thunderstorm begins

The early 1990s marked the dawn of a new service age. It was a true explosion of voice "supplementary services", CLASS services, information services such as airline reservations; and operational support systems, such as customer care centers. The technological maturity of intelligent network services and information services raised for the first time the question: "Can we create a business case by uniting the underlying ideas of IN and Internet services, choosing only the best approaches from both worlds and integrating them with flexible service management procedures?" This question was one of the issues tackled by the emerging TINA-C. Consequently, TINA started with a service- and business-oriented approach. Its uniqueness is that it clearly separates the service layer from the underlying network layer.

1.2.2 Driving forces in the service universe

In the last half-decade enormous changes have been initiated to prepare for the estimated $100 billion worldwide Internet-based service market in the year 2000. New brokerage architecture models are needed to shape electronic commerce. TINA-C's business model and the related reference points (cf. Chapter 6) represent an efficient frame for any type of service brokerage, retailer and provider activities.

A world of new services is now flourishing in the voice–data convergence area and in the fixed–mobile convergence area. In the voice–data convergence area, beside the "browse-and-talk", the "click-to-talk", etc., a fullness of possibility can be thought of; one example is the Web IN. To describe it in IN terms, Web IN would provide a number translation (done by the service control point) into the world of URLs, that would connect the SCP to a broader variety of services offered by the Internet. Dynamically created trigger points in service switching points provide a broader variety of services at a customer's disposal.

To illustrate another scenario, one could assume that for every incoming phone call, based on a caller's number identification, the latest case situation is simultaneously displayed on a screen. This time-saving procedure for attorneys, real estate agents and others is called computer telephony integration (CTI).

With all those tool sets, skills and excitement on hand, customers began to understand that there were different means by which they could play an active role in customization and control of their service needs (Figure 1.1). On-the-fly subscription, transparent and flexible billing procedures, and unified service delivery according to customer preferences are most popular when it comes to customer expectations.

There were still the notable differences in the quality of service paradigm, between the services being offered: "best effort" represented most significantly by the Internet or "guarantee of quality" always attempted by the telco community. The differences have often been ignored in the euphoria of service gold rush excitement.

1.2.3 Integration of mobile and fixed networks

As deregulation progresses, new opportunities appear for mobile and fixed operators:

- To integrate their operations into one unique customer service, and
- To offer the services developed for one network over another (e.g. IN service in the fixed network towards a mobile network; or terminal-oriented services on the

mobile network towards those on the fixed network, such as voice mail, message notification, and last calls journal).

The easiest way to develop fixed–mobile convergence is to apply seamless service portability by relying on a logical network connection management that will run on fixed–mobile, SDH or Internet network technology. Expectation rises for global seamless services: any service, everywhere, any time and supported by market players who ensure one-stop shopping or lifetime local number portability.

1.2.4 Multimedia for ever

What we are seeing as this book is being written is the shaping of the future information market: the new business opportunities and the variety of services which mirror the unrestricted diversity of people, their information and telecommunication needs. It varies through their goals, hopes and dreams, emphasizing certain features depending on their temporary role, e.g. a doctor on call, a secret admirer, a stockbroker or a video-on-demand retailer. The same intelligence must apply on both voice-over IP (VoIP) and on the data and video over IP connectionless services.

TINA architecture is explicitly constructed to serve this goal. New components are introduced into the network that are far beyond the elementary call and connection control functions of the traditional public switching networks for voice traffic. The principal conceptual extensions made by TINA in order to support internet multimedia applications are, for example:

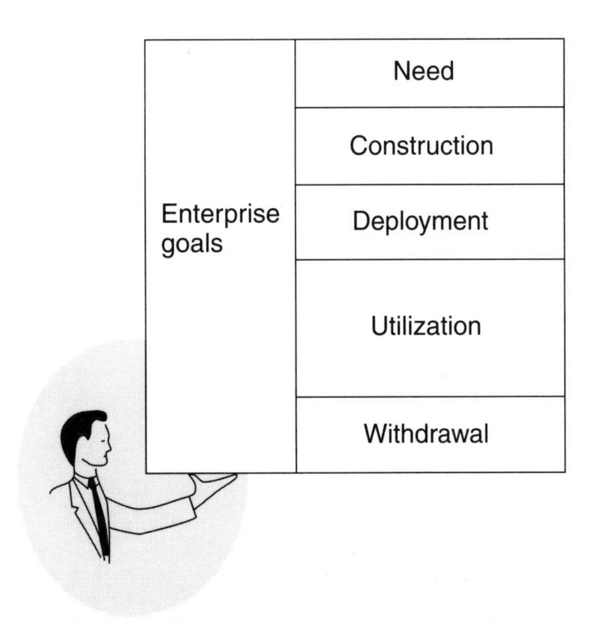

Figure 1.1 The customer wants to play a part in the service life-cycle

- Co-operation of several types of participants (stakeholders) such as service user, service provider, network operators to support a multimedia service. A TINA business model representing these elements is an integral part of the architecture
- Concept of communication session and service session
- Distribution provided at all levels of architecture, supporting evolutionary and technology-independent integration of new service functions.

1.2.5 Service management

This great variety of services has to be handled, controlled and managed. It is done traditionally (it is actually an important part of the telecommunications industry) by using a specific kind of service, so-called "service management". The distinction between services, network management services, and service management is going to vanish in the new service universe. This allows for real-time subscription, real-time provisioning of composite services, real-time changing of charging preferences and others.

The need for customer care grows with the new service complexity as well as with increasing competition. The credo "keep a customer, if you finally got him" requires new business models identifying roles (e.g. brokers, retailers or service providers) and allows us to break down the complexity of global customer care centers by designing them for each of these roles.

For completeness of this introduction, a more general question has yet to be raised in view of the business opportunities when dealing with service management. This is related to the authority to manage and control as a whole. Should management be centralized, distributed or both? Who is managing the telecommunications and information services and how is it done? Open distributed processing offers a lot of flexibility from this point of view by allowing various degrees of centralization/delegation.

1.2.6 Service creation

The Intelligent Network has progressively introduced service creation environments (SCEs) that enable quick customization or creation of services out of pre-integrated components (service-independent building blocks, SIBs). As competition between operators grows, there is an increasing need for differentiation in the service offer. This can be achieved by home-based operators developing part of the control logic or the operation logic, for example using these SCEs.

In parallel, the pressure to obtain services in local exchanges is coming to a climax because the complexity of software and system integration is near technical limits. In this environment, operators would like to share the risks of service development with the service providers using a more advanced service creation environment.

In addition, the expansion of services makes service interaction a nightmare; the combination of several services during a call establishment can produce unpredictable results (for instance, with call screening applied before or after call forwarding). As research on service interaction slowly progresses, the few results obtained converge on the benefits of sharing a unique service design, and a unique customer–agent and parameter basis in service composition.

1.2.7 Emerging technologies and market shares in service provisioning

Outstanding benefits are expected from an open service provisioning based on the convergence of telecommunications and distributed computing. Voice-over IP, i.e. Internet telephony, is a good way to illustrate one of those alleged threats to the "old telco world" or new challenging business opportunities (depending on the reader's viewpoint). An expected growth of Internet telephony service business is linked to improvements in voice quality and reliability of voice-over IP. As this threat increases, its impact on the long-distance voice tariffs will grow, until tariffs for voice-over IP and voice circuits become similar. Meanwhile, there is a temptation for the operators to invest in both markets, not knowing clearly what the future shares will be.

A large and still increasing number of national service providers exist all over the world. There is a tendency among them to build alliances based on the global service challenges to obtain a leading position in the market. Among them are Global One (with Deutsche Telekom, France Telecom and Sprint), BT and MCI, ATT and Unisource to name the currently most prominent companies.

As deregulation progresses, new opportunities appear for all kinds of operators to integrate their operations into one unique customer service, and to offer the services developed for one network over the others (fixed towards mobile for intelligent network advances, and mobile towards fixed for the terminal-oriented services such as ergonomy, message notification, and last calls journal): portability of services across various network solutions.

This global market test will call for solutions where operators may start cheap and small, and extend their coverage as soon as the service becomes popular.

1.3 TELECOM NETWORK EVOLUTION: REQUIREMENTS FOR A NEW ARCHITECTURE

The telecommunications industry has developed in the last decade a number of new solutions to handle the complexity introduced within the network by the growing demand for new services. These solutions are embedded in architectures which are a background for TINA, i.e. they offer inputs to TINA and need developments that are in the domain of TINA. The most important are without doubt Intelligent Networks, TMN and mobility.

One of TINA-Cs' (initial but still valid) objectives is to offer a migration strategy that will allow for a convergence of the Intelligent Network (IN) and the telecommunications Management Network (TMN), two recent developments within the telecommunications industry, to address two important issues:

- IN has been designed to provide a more flexible control of some telecommunication services by taking control of the switches where they used to reside (with all the engineering problems caused by the slow flexibility for evolution) and placing them at a centralized point.
- TMN has been designed to provide a more uniform means of managing hetero-

geneous network elements, from different vendors, that constitute the telecommunications networks.

Thus, one initial equation for TINA was "IN + TMN = TINA", meaning that both IN and TMN set requirements for TINA and that TINA is meant to deliver the convergence of the "control" world of IN and the "management" world of TMN.

1.3.1 Intelligent Network trends

1.3.1.1 Limits of a bottom-up approach

IN architecture[1] has been designed to execute remotely some of the switching control functions. The need to centralize the database for translation and routing in these services has driven centralized server approaches, where the servers (SCP) are accessed from decentralized transit or local switches (SSP). Typical examples of such services are virtual private networking or freephone.

The development of IN was based largely on a bottom-up approach in that the definition of SSP possible triggers and the existing basic call state model primitives preceded the definition of the SCP-to-SSP interface. At the start of IN, there was also a top-down approach in that an overview was made of the possible services and requirements for IN. But because this was done in 1986–8, the significance of broadband and the mobile multimedia services was underestimated.

As a result, the evolution of the SCP-to-SSP INAP interface towards the provision of multiparty and terminal mobility has been a long and difficult process. Each new release of INAP (releases are called CS1, CS2, CS3) include major architectural modifications of the interface, and thus of all the network elements, which makes deployment a major effort. In addition, the definition of INAP does not provide completeness so that interoperability between vendors in a specific release requires special bilateral agreements. We can summarize these as a requirement to evolve to:

- A more flexible interface to the switching networks
- Support for embedded mobility
- A more precise interoperability specification for the multivendor and multi-operator environment

1.3.1.2 Dealing with the emergence of multi-operator environments

IN network architecture introduces modeling in different planes: management, control and user. As this was conceived in the 1980s when the multi-operator environment was limited to a possible third-party service operator, these planes do not map to a business model many telecommunications operator roles. One could even say that they map only to the internal organization of one single operator: maintenance-TMN, network control, and terminals.

[1]See Appendix 3 for a brief summary of intelligent networks.

As a consequence, neither the INAP interface nor the management interface (nor the user-to-network interfaces), are fit to serve in an open multi-operator environment. None of the three can be easily protected for security purposes, nor do they provide a clean-cut separation between network and services, since each of them terminates message-oriented dialogs which are, for instance, a part of the call control.

In the same logic of mono-operation, the relation between the SCP and the end-user terminal was not structured, so that the switch could not organize a mapping between INAP orders and orders to the terminal, over the user–network interface (UNI), or over an application layer above the UNI. As a consequence, introducing the role of an IN retailer operator in this access part becomes impossible.

Additional requirements for evolution can be derived:

- New planes which map to business roles and permit a flexible multi-operator environment
- A clear separation of the interaction of the consumer role with network functions at all network levels.

1.3.2 Telecommunications management network trends

For a long time there has been a separation within telecommunication systems between the control logic (which has very stringent requirements on performance and fault-tolerance) and the management logic (which was developed over more mainstream solutions such as UNIX). This separation between control networks and management networks is a legacy situation due to several factors. First, management interfaces were extracted from the network element (in a physical alarm format, for instance) long before the same happened with the control interface (leading to the IN architecture). A second factor is that the management logic (that required less specific technologies, e.g. UNIX) could be developed by service companies and operators, independently of the telecommunications providers, who provided the real-time control logic embedded in switching.

There is a recent trend to mix control and management for at least two reasons:

- Their separation based on the technical differentiation noted above (e.g. "real-time" versus "UNIX") is no more obvious
- Creation and customization of new services is difficult (and sometimes dangerous!) if the introduction/modification of each "control" operation must be followed by a (separately developed) resource management operation.

It should be noted that IN has kept the management-versus-control duality. Even in the new IN CS3 architecture (currently being defined), a new business model for IN describes the different roles (among which those that an operator can play) and offers both a connectivity provider role as well as a management role that would require independent "management" operators.

This recent evolution has an impact on the telecommunications management network[2] (TMN) architecture itself, which is facing various issues that call for its evolution:

[2]See Appendix 3 for a brief summary of telecommunications management networks.

- The current TMN hierarchy in five layers, where only contiguous layers could exchange messages, is not flexible enough to serve dynamic service management; there is a need for direct access from the service management layer and customer network management to the network element (NE) or to the NE operation center, through non-hierarchical relations.
- These non-hierarchical relations have to rely on customized information screening (such as intelligent agent), and strict security checks, which are not yet available on the TMN.
- The continuous range of products from PCs to workstations call for software product portability across various architectures, for which object orientation and CORBA OO distribution can be a non-proprietary solution.
- Multi-operator co-operation at service management level and at network management level will become a reality in the near future.
- There is pressure to interoperate with or to re-use the private network management products, particularly those developed in the Internet community.

Globally, TMN evolution poses the following requirements:

- The integration of control and management functions and of control and management service creation
- Better support for scalability and durability of products
- Definition of co-operation for service management and network management that allow for multi-operator environments.

1.3.3 Mobility: universality and variety

Mobility will affect all types of network, because users will want to communicate anywhere efficiently and cheaply. Thus, mobility after the year 2000 will blur current boundaries between networks and services. Areas of overlap between different network technologies will be key competitive battlegrounds for operators.

The networks that are under design will need to cope with multidomain and integrated mobility issues. This is an easy basis for the introduction of TINA. We will discuss as an example the UMTS (Universal Mobile Telecommunication System), which integrates the previous mobile and fixed network solutions and provides new means to operate multiple broadband services. We could also have taken the third-generation Group Special Mobile (GSM 3G) as an example, as this evolution of the GSM towards broadband services implies a redefinition of part of the GSM architecture. In both cases, reference can be made to the conclusions of the Global Multimedia Mobility (GMM) European ETSI group (whose conceptual model is shown in Figure 1.2). It is important to note that the GMM model is based on *physically* defined interfaces: TINA has developed a set of reference points (see Chapters 2 and 6) that can easily be mapped onto the GMM interfaces to which they would provide more flexibility and multi-operator end-to-end openness.

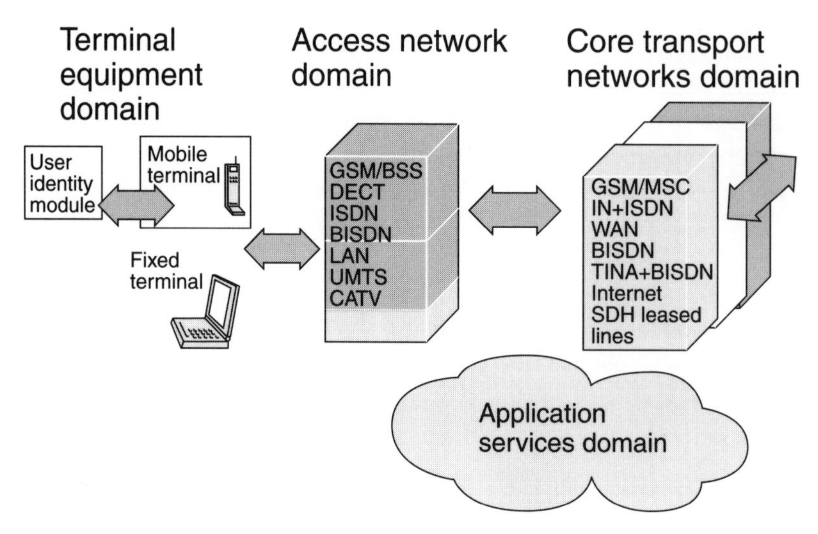

Figure 1.2 The GMM conceptual model of network architectures

1.3.3.1 Rationale for evolution

In general, the legacy fixed and mobile network architectures cannot easily solve the following issues in UMTS, GMM and global mobility:

- After deregulation, independent operators want to use their GSM core for entering new businesses: wired access and service offerings in corporate networks, for instance. This requires a flexible and powerful mobile core network, completely independent of a generic multi-usage access mobile network.
- Most major operators have both a fixed and a mobile operation network. After deregulation, they expect to merge the operations of both networks. More globally, there is a need for fixed–mobile convergence, such as combined mobile switching centers and local exchanges, that should reduce procurement and installation costs as well as the operational costs.
- From the user's point of view, there is a major need for service integration, such as integrated customer contract management and billing, and for combining fixed and mobile intelligence to always provide the best of each. This can be achieved by an increase in intelligence in the terminal, by co-operation of intelligence in the network, and by the standardization of common user agent, user session, and provider agent parts. This is done in TINA.
- There is a need for multiprotocol mobility, to cope with divergent protocols and interfaces such as GSM, DECT, CTM, CDMA, PHS and IN along with the multiple services markets offer.

This can be summarized in Table 1.1. The example of UMTS as a place to introduce TINA in legacy situations (in this case, the existing mobility and the next generation one) will be discussed in Chapter 7.

Table 1.1 Multiple services offered by markets

Mobility applications	Real-time communication Service Management Messaging Data Billing Management
Supplied to services	Freephone UPT Card call VPN Call screening Voice mail Personal agent
Accessed through terminals	PDA Multimode Pager Cellular Fixed phone Laptop
Connected to networks	PSDN Cordless Cellular Paging PAMD Satellite PSPDN
Using standards	ATM GSM pacsag NMT IN X.25 Mobitex ERMES (D)-AMPS GCPD ISDN

To summarize, there are requirements for any new service and network architecture, in order to support mobility in all its forms:

- Terminal mobility
- Service session mobility
- Personal mobility, keeping its context.

1.4 HUNTING FOR THE PERFECT ARCHITECTURE

1.4.1 What is expected from the candidates?

From the previous analysis emerges the picture of an ideal architecture that will satisfy a number of requirements: flexibility for business and technical change, overcoming some of the limitations of the existing architectures, and allowing a smooth migration path from legacy to the new services and networks. Will the reader be surprised if we assert that TINA is among the possible candidates?

Let us summarize the expected properties of an ideal architecture before making an early assessment of the possible candidates and their respective advantages. Keeping these requirements in mind will help the reader to better understand the presentation and discussion of TINA in the following chapters.

This "Architecture with Qualities" will:

- Support the current changes in the telecommunications business model as well as the changes due to enlarged deregulation and growing competition

- Support the rapid provision of new services and service features, at low cost and on a small scale, with strong scalability mechanisms
- Clearly separate service delivery from network provision and offer a means to port services over various networks
- Support multi-operator and multivendor co-operation and provide flexible and precise interoperability specifications
- Integrate control and management functions with a common engineering support
- Provide a seamless and universal support for mobility
- Accommodate intelligence "in the network" and intelligence "at the edges" and offer clear separation of the consumer from the network
- Define service creation principles and environments that allow production evolution and durability and incorporate advances in componentware and middleware
- Offer migration strategies that protect the installed network base.

Chapter 2 will present the choices made within TINA.

1.4.2 May the best one(s) win

At least four architectures are here to compete:

- PSTN and/or BISDN based on switches with embedded control and management
- Intelligent networks and mobile networks
- Internet networks (routers, point of presence/access points, powerful terminals)
- TINA-based services for BISDN, IN and Internet.

One (PSTN) has been around for a long time. Some were more recently deployed (IN and Internet). One (TINA) is the challenger and not yet deployed. Before going into the details of the challenger in the following chapters, let us examine some of their characteristics and roughly evaluate qualities of the four candidates. Table 1.2 stresses the need for evolution (including evolution of TINA in some cases since nobody is perfect).

TINA potentially offers means to overcome some of the listed limitations. The question of how and the migration strategies will be analyzed in detail in Chapter 7.

1.5 TINA: NETWORK EVOLUTION, NEW REVENUES, AND CO-OPERATION

A common thread to the evolution described in the previous sections is the central role of software. This is obvious for the information technology industry (it is enough only to recall that the software industry is now the third largest in the USA in terms of employees).

Table 1.2 A comparison between the main telecommunications architectures

	PSTN BISDN	IN/Mobility	Internet	TINA
Interface structuring principles	Signals	Messages	Frames, IP protocols	Objects, reference points, methods
Support to scalability	Limited by trunk signaling	Limited by INAP and MAP or IS 41	Excellent due to IT platforms	Excellent due to IT platforms
Separation of service and network	No	Yes, physical	No	Yes, logical (software)
Integrated control and management	No	No, management part is not standard	Yes	Yes
Efficient service-creation environment	Not yet	Yes, very quick as long as INAP fits the new service	Yes, terminal business	Yes, first investment may be large, but later re-use
Timeliness	Available	Ready, major investment made	Available but no network control	Mid-term
Shared development risk	No	Partly customization of prepared parameters	Yes (for applications and services)	Yes, can also imply third-party component offer

But it is not necessarily a new thing as well in the telecommunications industry: the introduction of software for the control of switching systems in the 1960s was an early revolution in the culture of this industry. Ever since, telecommunications systems have increasingly depended on the quality of software and, in particular, on its ability to evolve easily.

TINA is being developed (in particular within the TINA Consortium) with the primary objective of becoming a software architecture for services and for the operation of these services. This architecture should allow for:

- The rapid provision of new services or service features
- The reduction of costs, in particular through the re-use of components.

The architecture can also be applied to the evolution of existing services, and to the re-engineering of (or parts of) those services.

1.5.1 Improve the engineering of the telecommunications network

In the last thirty years, different developers of telecommunications systems have relied on solutions based on different business models and different technologies. The business models they have used are characterized by a very limited number of interactions between different administrative domains (even inside the various networks of the same public network operator) and by *ad hoc* resolution of the interoperability requirements. On the other hand, this has led to the development of many proprietary solutions and the gradual emergence of the "software crisis" severely constraining new product development. Different limitations are encountered in different parts of the telecommunications system: control of the network elements, management of the network and network elements, and computing applications such as billing. The following subsections in this chapter will investigate these limitations.

However, one of the common limitations encountered is the absence of good software structuring principles that will allow an easier introduction of new features or the re-use of previously developed "components". TINA has adopted from the very beginning the approach of defining generic modeling techniques that can be applied to any kind of services. Among the main solutions proposed are:

- The use of object orientation as the basis for TINA
- The use of distributed computing techniques as a means to allow for scalability
- The use of a flexible business model that adapts to various situations
- The definition of reference points that support business separations and risk sharing
- The end of the separation of control and management functions
- Others that will be described in the following chapters.

The TINA architectural principles (and the products derived from TINA specifications) will allow the re-engineering of network functions, offering a migration path for "legacy" systems (as described in Chapter 7).

1.5.2 Propose the conditions for the generation of new revenues

Re-engineering the telecommunications network is not the only objective of TINA. Even more important is to address the provision of new services.

One of the main objectives for the work of TINA-C in the last few years has been to consider the complexity of the provision of multimedia services. These services cannot be provided without the use of (sometimes complex) means to co-ordinate the resources that are used by the participants in such a service: TINA has defined a notion of session that addresses this complexity. Another source of potential new revenue, and an important objective for TINA-C, is the application of TINA structuring principles to the provision of information services on top of the Internet that support flexible business roles (to accommodate the ever-changing nature of "Internet business") and offer guaranteed quality of service.

In addition to its architectural principles (e.g. components or reference points), TINA offers a variety of mechanisms to support the exploration of new services on small prototypes and ensure the scalability of the solutions. These aspects will be detailed in Chapters 3 to 6.

1.5.3 A co-operative solution for a competitive world

It is likely that the Information Society will be based on a very solid telecommunications ground and a telecommunications network of much greater complexity than the one we have today: an information network. It will also be the result of the (collaborative) competition within this sector and other sectors such as the information technology industry and the content providers. Deregulation will allow lowering of the "entrance barriers" and provide access to new business entrants.

In the process, the role of standards (*de facto* and *de jure*) remains central. To distinguish oneself from competitors by introducing new services or service features has been proven as a successful way to obtain a forerunner position in the market. Unfortunately, this often meant in the past running far ahead of standards as well. Therefore when the slow standardization process had finally been finished, the early introduced products and services were possibly not in line with the later upcoming standards. Since a slow introduction of new services to the market is no longer an option, other business approaches are needed. It can be either to dominate the market with the earliest leading introduction of a new solution or to form quasi-standardization bodies in order to obtain early agreement on which way to go before fitting it into a standardization process. Whereas the first approach is risky on complex implementations, there are several groups and fora today following the second path. TINA-C itself is an excellent example of unifying the interests of different equipment and computer vendors with large telecommunications companies worldwide for the benefits of the newly shaped information marketplace.

Winning solutions will be those that offer the maximum flexibility (to deregulation, changes in business organization and in network technologies, etc.) together with a solid common basis on which business differentiation can be made. TINA-C has implemented from the very beginning the means to provide the common solutions:

- TINA makes use of as many existing industry standards or results as possible. TINA-C's relationship with the OMG for the computing platform is a good example (TINA is based on CORBA and focuses only on the additions to CORBA).
- TINA specifications have been defined by a consensus approach and are open to the public.

TINA-C, which is a not a standardization body in itself, advocates the adoption of its results by various standardization bodies.

Chapter

2 What is TINA?

2.1 INTRODUCTION, FLEXIBILITY AND OBJECT ORIENTATION

Having identified these new business opportunities, TINA has been designed to provide an integrated service management capability whatever the information and/or telecommunications service. This can be seen as one of the prerequisites for offering flexibility in an open information marketplace. Everyone and everything, acting in the business, can manage or can be managed. TINA provides greater flexibility than the TMN. It allows dynamically changing structures with regard to stakeholders and their roles, domain borders, ownership and price structures.

The TINA-C architecture proposed solutions to these issues along the lines covered in this book. As they are designed using object orientation, TINA-C service components can be flexibly initialized at their creation at a service factory and later enhanced with FCAPS functionality as desired. We will further explain how service federation and composition in a multibusiness domain environment is contained in TINA-C.

The vision of the expansion of services has already been outlined, but the technological choices have been left to software, IT systems, and telecommunications engineers. Nevertheless, it has to be taken into account that all those who are building the future service business are in a fast-changing environment which determines to a larger extent the success or failure of their service in the marketplace. Some of the most influential facts are highlighted below.

TINA-C methodology allows services to be built in a well-designed manner by separation of concerns. Based on a universal service component model (USCM), object grouping, packaging and composition guidelines, the re-usability of service components is one of the prominent characteristics of the architecture.

TINA is an integrated solution, as opposed to IN architecture, including a top-down approach. That is, before designing the general TINA architecture, a complete expression of the services that it should support was assessed and the architecture was based on these requirements rather than a bottom-up assessment of existing capabilities of network elements.

The new service architecture was defined to have a clear independence from the underlying network elements connection control and management which was defined in a network architecture. Additionally, the TINA reference points were introduced to make clear differences between the business and technical separations between actors in a multi-enterprise service and a network environment. For example, the aim was to make a TINA network easily opened to new third-party service operators; the interface between these services and network capabilities being presented in the ConS and TCON reference points is described in Chapter 6.

The TINA interfaces are not based on "call model" signals and triggers, they are object oriented; this provides a flexibility to enrich objects and methods without major changes in the element kernels. Because TINA interfaces rely on information modeling and object orientation, the service logic in a server of the network can be upgraded without impacting the releases of the resource network elements. Of course, this assumes that the information model designed at the start of the project is sufficiently complete to encompass the new usage by the upgraded server. For example, the introduction of performance management in a network where only fault management was active previously requires an upgrade to the information model of the interface.

TINA seems to solve efficiently most of the previous issues of the switching and IN architectures related to multi-operator service offers. Nevertheless, introducing complete TINA flexibility into the network also implies the introduction of new interfaces into network elements. Because the modification of public networks has a major investment cost and a long time to market, the introduction of TINA requires the use of migration paths, relying on legacy interfaces and signaling. This is detailed in Chapter 7. Intermediate migration steps will probably keep legacy signaling interfaces (IN CS1, for instance); this means that current issues like the lack of "clean functional cuts" may persist in intermediate migration steps.

2.2 TINA – FRAME OF REFERENCE

2.2.1 Origin and evolution of TINA

At the start, the TINA Consortium was set up to achieve several goals:

- A solution to improve the way services are designed
- Application of the TMN paradigm to the IN architecture
- A way for the operators to develop services on their own in open multivendor computing and service platforms.

During its development, TINA was assigned a moving target as each stakeholder wanted to apply the new technology to the most promising market segments. It moved from multimedia (cf. DAVIC), onto the ATM Forum service management, to the Internet value-added services, and to the intelligent networks. The Consortium Core Team was mixing young engineers from the major companies in telecommunications and computing; it happened to be a good place to check each other's comprehension of the market, and to elaborate future service specifications.

The novelty of TINA was that it solidified emerging telecommunications ideas about the inherent value of service content above the hitherto raw supply of connections that left content up to end-users. Many ideas were drawn from previous research programs such as ANSA and the European RACE project Cassiopea. TINA's initiative was to bring these ideas together in 1993 into a telecommunications, computer networking and software forum.

The basic ideas were formed around a traditional telecommunications business model with consumers, providers, network operators and well-anticipated requirements (such as consumer mobility and ubiquitous access). Work progressed using ODP perspectives resulting in information modeling, object specifications and abstract computational ideas that had not yet resolved many distribution engineering issues. During TINA's second year (1994) a number of TINA auxiliary projects started, as described in Chapter 8, with the aim of validating the architecture through prototyping and implementation. Such projects encountered difficulty in choosing appropriate implementations of the architecture because object distribution and enterprise responsibilities for information and computational objects were too immature.

Later refinements of the service architecture and emergence of the CORBA distribution standard were visible in 1995, as well as pressure on the trader as an inheritance of previous DCE and ANSA activities.

Later stages of the service architecture (SA) definition in 1996–7 were influenced by the rapid growth of the Internet – which moved from rudimentary text browsers to the multimedia browsers and Java during that period – making it an obvious first TINA deployment opportunity. The SA teams responded by ensuring that its enterprise paradigm was consistent with the Internet, and positioning the SA to add control and management value on top of the Internet. Additionally, specification of interfaces was accelerated since these represented the real source of service level interoperability, embodying in CORBA IDL the engineering optimization of the information and computational specifications. It was in the same years that the reference points were specified, and that the roles of the various operators at the point of presence were detailed.

This summarizes the various waves of TINA and industry concepts. Documents of the various years are available on the TINA server at www.tinac.com. Work on consistency in the second set of TINA-C specifications builds the reference TINA architecture.

Many early TINA concepts and documents, as befitting research, were rather theoretical and their implementation frustratingly elusive. The aim here is to cover some of the practical and interoperability issues of primary concern to designers, so we will describe mainly the latter parts of the Consortium's output, which build a coherent set of specifications.

2.2.2 Requirements

The first requirement is to provide a common architecture for the provision of services, unlike existing telecommunications architectures which play on overlays. This will enable consumers integrated access to a set of customizable services, independently of the network offers of operators.

The second requirement is to hide the complexity of the existing network in order to enable operators to rapidly introduce new services within a service infrastructure. This will thereby reduce operations and maintenance costs. Furthermore, the generic TINA computing platform has to

- Ensure portability across multivendor equipment
- Enable operators to develop their own services
- Make it possible for them to benefit fully from advances in computing technology (e.g. object orientation and distribution)
- Offer independence from telecommunications manufacturers' product lifecycles
- Guarantee easy interworking between operators.

The use of a common architecture for all kinds of services is a requirement that will enable manufacturers to adopt a common approach to services and their management: reallocation of functional entities in different architectures and better integration of applications. It will, in the long run, reduce development effort and improve software quality as a result of distributed processing.

2.2.3 Different aspects of the TINA architecture

An architecture aiming at capturing the complexity and rapid evolution of the world of emerging services cannot be designed as a monolith. Even if TINA-C has attempted to address the largest set of issues to make the architecture as complete as possible, it has never been TINA's goal to be a "big bang" replacement of existing services and networks. TINA brings a choice between the following:

- Software engineering: a way to introduce in the telecommunications industry a set of modern concepts (OO, distributed computing) in order to allow for an upgrade of the quality of the existing products.
- Platforms: a way to introduce large-scale distribution and interoperability; and the corresponding tool-chain.
- Business: a new way to define the roles in the industry and to propose methods to play these roles in a more flexible way.
- Services: a new way to separate them from evolution of network, to structure their control by session.
- Networks, including the Internet: a description of new networks based on different approaches to control (e.g. third party).

- Design: a new approach to service specification, new service methodologies, the evolution of the concept of SCE.

2.3 MODELING TINA SYSTEMS: BUSINESS MODEL

2.3.1 Overview

The basics of object oriented design and analysis show that one must use several viewpoints in order to define the aspects of a system in a complete way. A classical reference for this analysis is the use of the RM-ODP architecture, which defines five viewpoints:

- The enterprise viewpoint specifies the roles of the external actors of the system
- The computational viewpoint specifies the abstraction model of the system, without knowing how it will be implemented
- The information viewpoint gives the data model, inheritance links and relations between object lifecycles
- The engineering viewpoint gives the system decomposition, taking into account the implementation on a defined target
- The technology viewpoint gives the system implementation up to the physical levels.

The ODP approach has been basically adopted by TINA, though the TINA architecture is not totally compliant with the ODP standard. Indeed, some customizations of the ODP model have been introduced, especially in the building of the TINA service architecture and network resource architecture.

To summarize, the main features of TINA are:

- A business model to be used as a general framework in order to determine the reference points and related interfaces for the services
- A software architecture based on the object oriented approach, following RM-ODP
- An application architecture targeting the telecommunications domain, that is, mainly the service architecture and network resource architecture.

2.3.2 TINA business model

The new telecommunications and information networking services will only be of value if they can follow the dynamics of today's consumer market. Services need to be rapidly created, introduced and provided. Also the management and administrative processes need to be in place. This means that TINA needs to address not only the services and systems but also the administrative side of providing TINA services. Additionally, the market will become increasingly complex and the players in the marketplace will change frequently and rapidly. Business relations between players will have to be able to conform to the same dynamics as the TINA services. Thus the TINA Consortium has developed a business model that is capable of the required dynamics.

The TINA business model provides a framework that allows stakeholders to establish and terminate business relationships. The concepts introduced can be used as a means to quickly establish bilateral agreements within a TINA framework or to establish global "ITU-T"-type standardization and conformance. It utilizes the concept of reference point as a point of conformance to TINA specifications and provides the means to dynamically establish and manage these. In Figure 2.1 an overview of the concepts of the TINA business model is given.

The basis for operation of any TINA service or (sub)system is the interactions between the informational, computational and engineering objects owned by the various stakeholders. Thus the concept of the separation of these stakeholders was introduced. Stakeholders are separated by reference points into business administrative domains. Interactions between the stakeholders are specified in reference points that are constrained by a business context called a contract.

These mechanisms can be generically applied to any type of stakeholder interactions whatever their business structure. However, to fulfill the possibility of global interoperation of TINA stakeholders a minimal common structure needs to be introduced. The TINA proposed structure contains a consistent set of five business roles based on the existing businesses in the telecommunications and Internet marketplaces. The interactions between these business roles are described as business relationships.

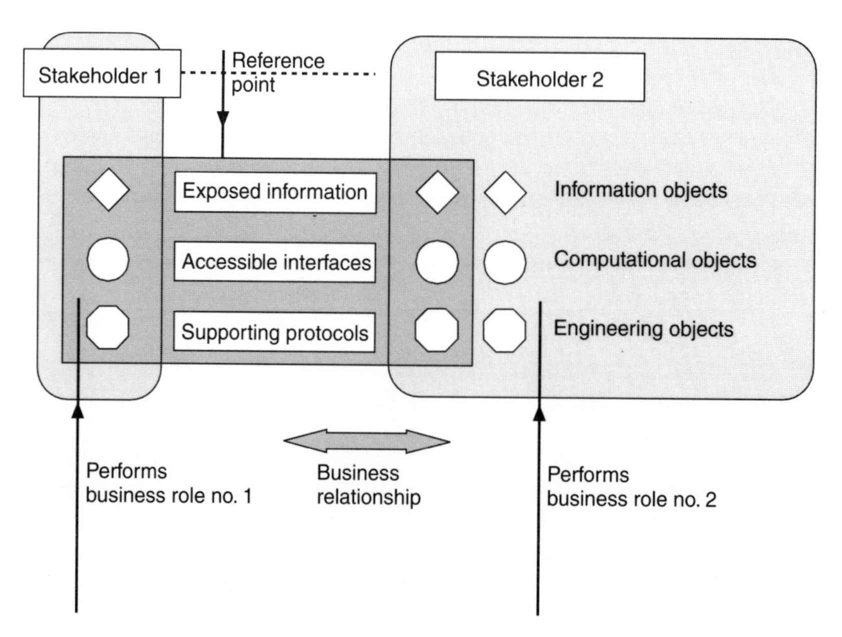

Figure 2.1 Overview of concepts in the TINA business model

2.3.3 Initial business roles

The following types of business roles (see Figure 2.2) are identified for the initial set in TINA:

- The consumer business role is introduced through economic considerations, as it is the only business role "consuming" the TINA services and not trying to make money from them. All other types of business roles are characterized as either "producers" or "middlemen".
- The retailer is separated from the third-party service provider business role through economic as well as technical considerations. Using a supermarket analogy as an example, the production of services (e.g. a movie) requires a different business set-up and technical skills than the offering of this movie (e.g. a video shop). The retailer business role is oriented towards customer management and value adding, while the third-party service provider is oriented towards production and maintenance of the service.
- The broker business role is introduced through regulatory considerations to allow all consumers to have fair and equal access to information about how and where to find services and business administrative domains in the TINA system.
- The connectivity provider business role is introduced through technical, regulatory and commercial considerations, since the lifecycle of new transport resources (including hardware and software) will be different from the lifecycle of the predominantly software service supported on those networks.

2.3.4 Business relationships and reference points

The business relationships can have varying lifecycle speeds and impact. On the one hand, the relationship between consumer and retailer can be transitory where in a matter of minutes the relationship is set up, a service requested and provided, and the business relationship terminated (e.g. making a call from a payphone). On the other hand, the business relationship can be long-lasting (e.g. a domestic telephone subscriber) and all-encompassing (e.g. one-stop shopping relationship).

A reference point is a specification of the requirements posed by the business relationships it implements. It will consist of several viewpoint-related specifications governed by a contract. As such, it is an aggregation of these specifications and has no direct meaning in relation to a business administrative domain.

Between the five business roles described above, the interactions (business relationships) shown in Figure 2.2 are envisaged. Among these relationships some have been considered as more important from the point of view of TINA's implementation. This means that a particular emphasis has been placed on the specification of the corresponding reference points. The three major relationships are:

- Retailer business relationship (Ret) providing access and management to end-user services and lifecycle management of users

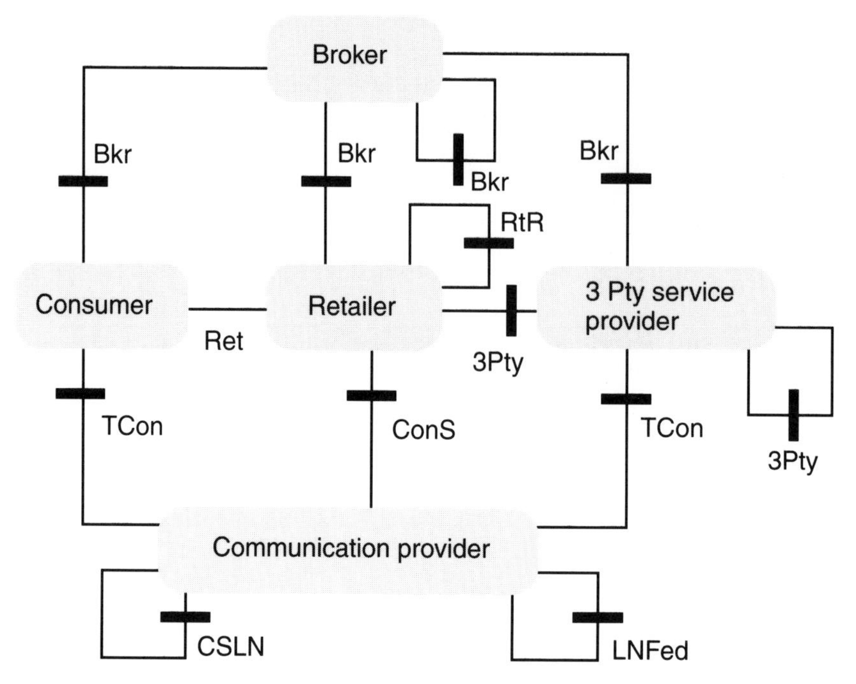

Figure 2.2 Initial set of business roles and business relationships

- Connectivity service business relationship (ConS) providing control and management to connectivity
- Terminal connection business relationship (TCon) providing access management between the connectivity provider and the connectivity user

These reference points will be described in more depth in Chapter 6.

A second set of reference points intended for longer-term specification in TINA-C were identified because they provide interoperability between different service providers and federation between connectivity providers:

- Third-party business relationship (3Pty) providing access and management to content and service logic
- Retailer-to-retailer business relationship (RtR) providing access and management to services
- Layer network federation business relationship (LNFed) providing peer control and management between connectivity providers
- Broker business relationship (Bkr) providing exchange and management of the broker information

- Client–server layer network business relationship (CSLN) providing control and management of the server layer network.

In summary, TINA reference points fit with the usual telecommunications idea of "interworking standard interfaces". Some are interfaces between different roles (TCon), others are interfaces between operators of the same role (CSLN). Until now, very little effort in the telecommunications standards had been invested in the interfaces between different roles. The first objectives had been to run network interfaces between connectivity operators.

Many questions were raised on the sensitivity of this model and the introduction of new roles, for instance the "information content provider". So far such extensions have been considered as of the third-party role and corresponding reference points; 3Pty is specialized for the new role.

One should not consider the TINA business model as the ultimate and unique model for telecommunication business. Its application to Internet, for instance, requires a few changes in the role of the network connectivity and service provider, since there is somehow two different levels of network connectivity and of service and content.

A business model similar to TINA's was elaborated by ITU-T with the same five roles, but adding a new role, "network management operator". This role was built out of the previous IN functional view, to extract network management aspects from all the information models of the previous reference points. This finally causes a split in two of all previous reference points: one being their "control part" and the other their "management part". The TINA object oriented model benefits from integrating the management and the control operations of the same service elements. When the two aspects are separated, there is a good chance at the service creation phases that the information models will become incoherent.

In conclusion, the TINA business model is a general framework useful to identify the appropriate roles and reference points of a family of telecommunications services. It has to be instantiated for a specific family of services. One should complement the network interface definition by the service component interface definition to permit software from various vendors to interwork at the reference points. These should be considered as guidelines for the development of components and at this stage, the reference points cannot be considered as a detailed TINA compliance specification.

2.4 TINA MODELING: SOFTWARE ARCHITECTURE

2.4.1 Overview

What is intended by the software architecture is the result of the OO analysis and design activities which cover both the computational model at the most abstract level and the engineering model when mapping to a precise distribution model is needed. So a TINA object can be detailed in its computational object (CO) view or in the engineering view of its computational objects, called eCO. Both views of this object refer, of course, to the same part of the information model.

2.4.2 Information model

The two kinds of modeling notions (with partial tool support) used by TINA information specifications are:

- The object modeling technique (OMT) for graphical representation, now migrated to the UML object representation
- A variant of GDMO (caller quasi-GDMO or q-GDMO) is used for describing the information objects.

In that sense, TINA object modeling uses only market tools and representations. It has stressed the need, beside UML, to have a language for describing object behavior which may be the majority of the implemented code. In many TINA application projects, SDL was used for the behavior specification. For the interface specifications, IDL is extracted from the well-structured GDMO templates in later phases of the software development lifecycle.

Two notations are used within TINA-C documents, UML more than q-GDMO (which is mostly used in the network resource information model, NRIM, see section 2.5.4.1). However, neither is prescriptive. UML corresponds to the mainstream notation in

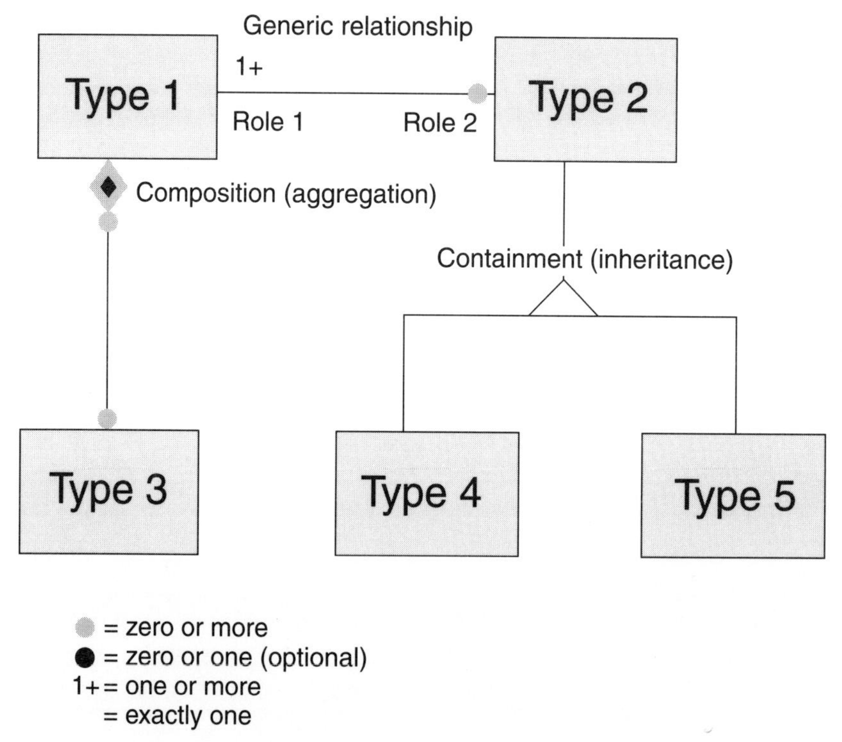

Figure 2.3 Example of information modeling with UML

object oriented analysis and is supported by a number of tools from the market. The need for rewriting GDMO specifications from legacy specifications into q-GDMO makes the use of q-GDMO limited in the current version of the architecture.

2.4.3 Computational model

It is useful to note that the TINA computational model is based on the conceptual distinction, fundamental in the object oriented paradigm, between type, which is subject to specifications, and instance, which is subject to "life" (creation, deletion, etc.); an instance is of a specified type. This distinction applies to COs, CO groups and interfaces. A potentially distributed application consists of a collection of computational objects (later called objects), which can be gathered into object groups.

2.4.3.1 Objects

Object-encapsulated data (or state) and processing (with behavior) provides a set of capabilities that will be used by other objects through its own computational interfaces, which are the interaction points for other objects: there is no other way to access the services or data inherent in this object. Some of the interfaces provided by an object may be offered when the object is created while some may be offered dynamically (i.e. created) during the execution of the object.

An object can offer several of these interfaces. This diverges from the OMG first ORB model, where all the methods of an object are grouped into one object interface. Why did telecom applications require this detail between different object interfaces? The main reason is the use of application management or configuration interfaces to complement the on-line control interface between objects. The management interface of an object would be used, for instance, for setting and upgrading part of the data and behavior of the object and the configuration interface for the object lifecycle.

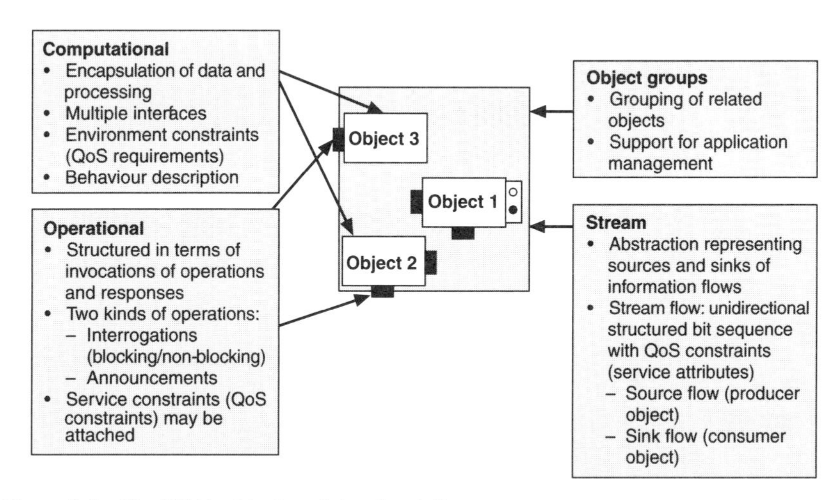

Figure 2.4 The TINA object model and notation

Object interfaces are called operational interfaces when they deal with the objects' interaction. Others are stream interfaces, for which the object does not interact with the information flow carried over the interface: see Figure 2.4.

2.4.3.2 Operational interfaces

The interactions which occur at an operational interface are structured as a set of invocations of one or more computational operations (operations for brevity) and responses to these invocations. At a defined interface, the object that invokes operations is called the client, and the object which provides the response is called the server, in the traditional client–server model. An object can be both client and server in its interaction with another object. In that case two different interfaces are used (see Figure 2.5).

There are several types of operations that can be invoked at a TINA interface:

- Announcement: this operation consists of passing zero or more arguments from client to server and processing a following action in the server. No answer is expected from the client: no result is passed back, nor any status on the processing by the server.
- Interrogations, blocking interrogation: this operation consists of passing zero or more arguments from client to server, then processing the invocation at the server, then returning zero or more results from server to client.
- Non-blocking interrogation: the same operation, but the client does not wait for the answer: the result is processed asynchronously when received.

Service attributes may be attached to an invocation. They include quality of service parameters.

Objects execute concurrently with respect to one another. The activity structure within an object is not specified in this interaction model. In particular, an object may process concurrently several operation invocations and responses, and thus may have

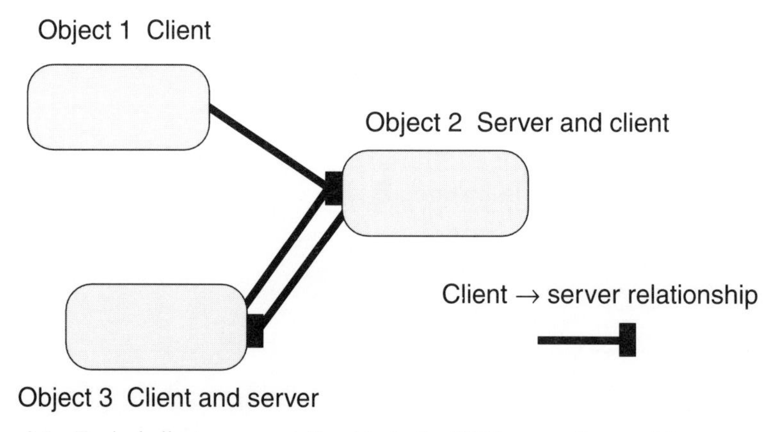

Figure 2.5 Typical client–server relationship in the TINA computing model

several concurrent activities within it. Concurrent use of an interface for file access may be allowed, while concurrent use of an interface for audio communication, for instance, may not be allowed.

2.4.3.3 Stream interfaces

A stream is an abstraction that represents a communication for some information flow in which content is ignored. In OMG, this stream is visible only through its control: e.g. for a video recorder, you can address only the operations "play, stop, fast forward". In TINA, a stream flow can also be controlled but, moreover, you can set up the stream channel through configuration, and some flexibility is introduced in the control of the stream such as negotiation of the quality of service.

Technically, a TINA stream interface is the computational manifestation of a stream flow end-point group (SFEP group in the OMG terminology). A stream interface supports data control functionality for associated SFEPs and stream flow connections (SFCs).

A TINA stream flow is always unidirectional, and is organized as a sequence of bits which might present a frame structure at the flow application level; quality of service parameters which are notified include timing requirements of frames and synchronization requirements between flows. A TINA stream interface is the termination: it can be a source or a sink for the flow. In order for two objects to interact by means of stream flows between them each object has to offer a stream interface and an explicit binding of the two interfaces must be realized. The stream interfaces are used not for the mapping of computational objects but mostly to model the telecommunication flow (voice, video, etc.) upon which the TINA objects build control and management services.

2.4.3.4 Object groups

An object group is an abstraction which supports the hierarchical aggregation of objects. The main motivation for object groups is to ease the management of a collection of objects: for instance, to load together, or to relocate together a collection of objects by a single operator. The concept of object grouping can be a vector to support multiple levels of application structuring.

An object group can be organized with a group manager object, responsible for managing the objects within the group (*vis-à-vis* the start-up, for instance). The group can be considered as a single object visible through the group interface. As opposed to a single object, the visibility of the internal structure of an object group is known only from the group manager.

In the following, a TINA application is called a service component. The functional view shows the object classes and the structural view the object instances architectured in the computational model. Further details of implementation would be visible if the engineering model was presented.

2.4.3.5 Service components

At this point, it is necessary to define a way for service components to be specified. It is very important that a service component has a unique, well-defined specification, also independent of the component internal structure. ODL is the language to be used for this specification, since it is the TINA language for computational specification (see Chapter 3 for the ODL presentation).

The concept of service component captures a functional view of the TINA service architecture, in that a SC can be specified without prescribing its internal structure. The multiplicity of CO mappings for a SC guarantees the flexibility of the structure, while the uniqueness of the CO representation ensures unambiguous specification.

2.4.4 Engineering model

The engineering model of TINA is presented in detail in Chapter 3. It maps the engineering objects of the computational model described above to the distribution infrastructure DPE through an abstract distribution model.

The distributed processing environment (DPE) is based on an object request broker compliant with the OMG specifications; this ORB is complemented by DPE servers which enrich the distribution with elements necessary for the telecommunications applications: notifications, transactions, etc. At the same time the interface definition language (IDL) is enriched to cope with the telecom extensions and to describe the object internal logic as well as the object grouping. It becomes ODL, which has now been adopted by ITU-T for the telecommunications services specifications. The abstract distribution model uses object clusters that communicate between themselves using stubs, channels, binders and possibly protocol adapters.

2.4.5 Assessment of situation

It is not mandatory to apply all the previous analysis and design methods to obtain a fully TINA-compliant architecture. Here in decreasing order of importance are the benefits of TINA in the different phases of the development lifecycle:

- TINA has specified in detail the computational model which helps in using the object distribution in a transparent way well adapted for telecommunication usage.
- TINA has specified an original business model, which is the reference in telecommunications for multi-operator co-operation and federation.
- TINA has specified a model information model that has been standardized in the ATM Forum.
- TINA has not covered the engineering model in detail, as this part can be left dependent on the target platform.

2.5 TINA MODELING: ARCHITECTURAL FRAMEWORK

2.5.1 Overview

TINA architecture enables pieces of software for a specific service to be distributed over the network and allows new types of services that are only possible through close interactions between software components on the terminal and those in the network. TINA is based on a common software components model applied to all parts of telecommunications information systems, including terminals (personal computers, etc.), transport servers (switching systems, routers, etc.), service servers (VoD, Web, etc.) and management servers (authentication, billing, etc.).

This software model eases complex development and deployment between different stakeholders by allowing distribution of software components over different blocks in the network and by decoupling them so that the evolution of the underlying technology of one component (standards, languages, programs, materials, networks, etc.) does not affect other components. The software components will be portable therefore reducing the cost and time of their development.

TINA provides precise separation principles. A major separation is made between telecommunications applications, on the one hand, and the environment on which they run, on the other. Telecommunications applications themselves are again separated into the part dealing with the provision of service and that dealing with the control and management of the resources needed.

These separations are reflected in the three main components of TINA architecture. (Some are prescriptive, namely defining a set of properties to which TINA systems must conform. Others are simply descriptive.)

- The computing architecture describes the distributed processing environment (DPE). The DPE can reside in multiple heterogeneous pieces of equipment and make them look like a single system for the software in the applications and middleware layers. The TINA DPE is an extension of OMG's CORBA adapted to telecommunications systems.
- The service architecture describes a universal platform for providing various services in an environment supplied by multiple vendors. It is based on a notion of session that offers a coherent view over a certain period of time of the various events and relationships taking place in the provision of the service.
- The network resource architecture describes the control and management of network resources. One major target of this architecture is to model in a common manner the control and management functions that are usually separated in current networks. In addition to this principal goal of unification, the NRA has been designed to take into account the complexity of the connectivity requirements of emerging services such as multimedia that need to accommodate complex communication sessions and to guarantee quality of service. The NRA uses a "TINA network" model derived from a new hierarchical connection management and presents a unique point of control. This architecture has been specified in detail (and has been the object of prototype implementation), but it is descriptive. An implementation, especially on existing networks, can use other principles

2.5.2 Computational architecture

The primary role of the DPE is to hide from the service creator the nature of the underlying networks, notably distribution. In ODP terms, the DPE assures a certain number of "transparencies" (e.g. distribution). An application will be described in the form of generic and specific computational objects interacting over the DPE in a location-transparent manner.

The DPE function provided over a series of DPE nodes may be differently implemented (a typical implementation is based on CORBA 2.1) linked by the kernel transport network (kTN). The kTN is a logical network and can be implemented in a number of ways (LAN under TCP/IP, etc.).

The TINA DPE is based on the communication services of the CORBA ORB and on the common services that can be compared to CORBA Object Services ("trader", persistence, transaction, lifecycle, security, etc.). Moreover, it offers extensions to CORBA that are specific to telecommunications such as "streams" (continuous information flows between two computational objects), the possibility of offering multiple interfaces for an object (for example, one for management and one for configuration) and of expressing information about quality of service (QoS).

These extensions, relatively limited and clearly delimited, are worked out within the special telecommunications group in the OMG, and some of them are on the way to adoption. The maximum alignment of the DPE specifications with those of the OMG is an important objective of TINA-C.

2.5.3 Service architecture

Service architecture is the newest and the most important part of TINA specifications and is based on several fundamental principles (see Figure 2.6):

- Separation of objects into generic objects (common to all services) and specific objects (representing service logic, data, management, etc.)
- Use of the notion of session that represents the information used by all the processes involved in the provision of a service for certain duration. For example, in certain services such as an audio/video multipoint conference, information about the multipoint connections and charging or the user's profiles must be retained during the duration of the conference, during which participants of the conference may come and go. Such sessions can be very complex as well as very simple (e.g. search of a web page).

Three main separations are defined in a TINA session that correspond to different types of activities taking place:

- The access session corresponds to the establishment of the terms and conditions of the session (e.g. authentication, selection of service profile, etc.) during the connection of a user to a system. It will allow the user to start service sessions, combine sessions, and become involved in several services.

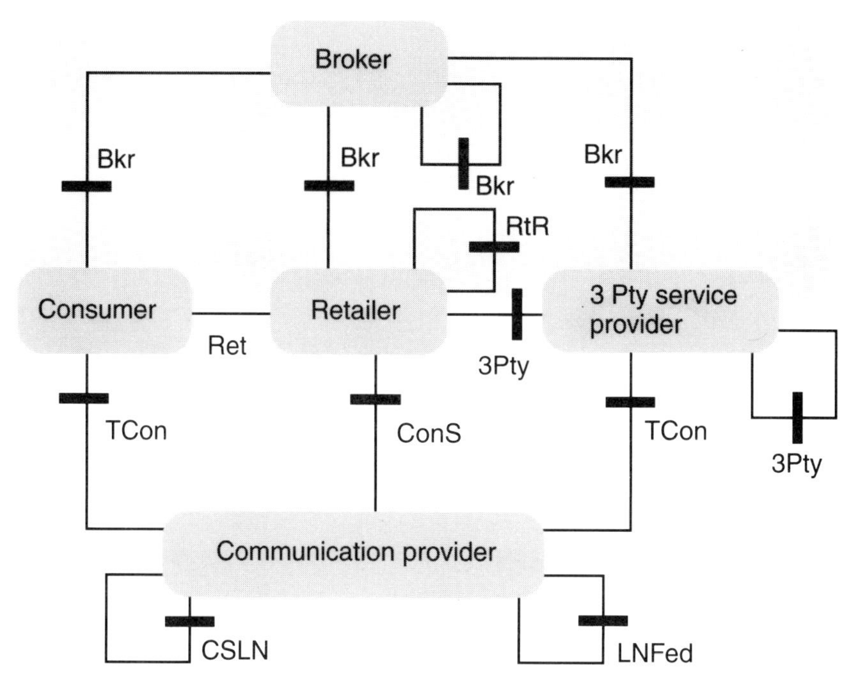

Figure 2.6 Scope of service architecture

- The service session corresponds to the provision of the service itself (e.g. moderated multiparty conference with additional information retrieval capabilities, etc.) implying the control and management of the overall coherence as well as of the users. It is separated into:
  (a) A service user session that manages the state of each user's activity and resource attributes (e.g. charging context, current page, etc.)
  (b) A service provider session that contains the service logic and offers the functions allowing the user to join a session, to be invited to a session, etc.
- The communication session corresponds to the co-ordination of the network resources used. It provides an abstraction of the actual connections in the transport network (communication routes, termination points, QoS, etc.) (see Figure 2.7).

2.5.4 Network resource architecture

The main target of network resource architecture is to describe the control and management of network resources in a common manner, whereas the control and management functions are usually separated in existing networks. In addition to this principal goal of unification, the NRA has been designed to take into account the complexity of the connectivity requirements of emerging services such as multimedia that need to accommodate complex communication sessions and to guarantee quality of service.

The scope of the NRA in terms of business roles is primarily to describe the con-

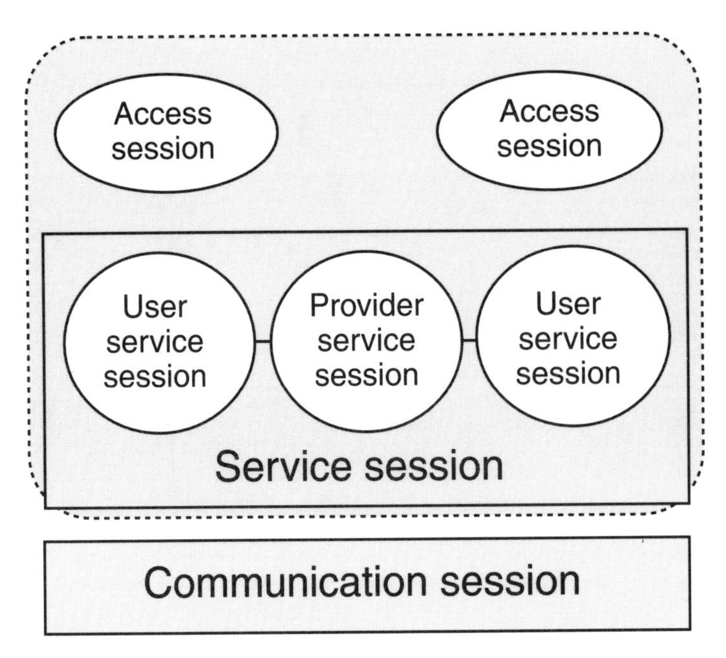

Figure 2.7 Scope of service architecture: sessions

nectivity provider role and to do this in a technology-independent manner. However, it also deals with the roles of the client of the connectivity provider (consumer, retailer and third party in the current definition of the business roles) (Figure 2.8).

To speak in terms of TINA sessions, the scope of the NRA is also to describe how the association between end-points and network resources is made and how the communication requirements of the service session are met. This is the role of the *communication session* (Figure 2.9).

2.5.4.1 The network resource information model

The NRA uses a "TINA network" model that is based on the TINA network resources information model (NRIM) which is a specification of the information elements that represent the topological and connectivity structure in such a network. The specification is technology independent and the control and management functions can also be derived in that way.

The NRIM addresses only a part of the functional management areas with a particular focus on *connection management*. This connection management is intended for the management of a variety of functions from a very high level (management of stream flow connections) to a very low level (at the network element level).

2.5.4.2 Service view on connectivity: connection graphs

In order not to give many (unnecessary) details on how the connectivity is provided at the

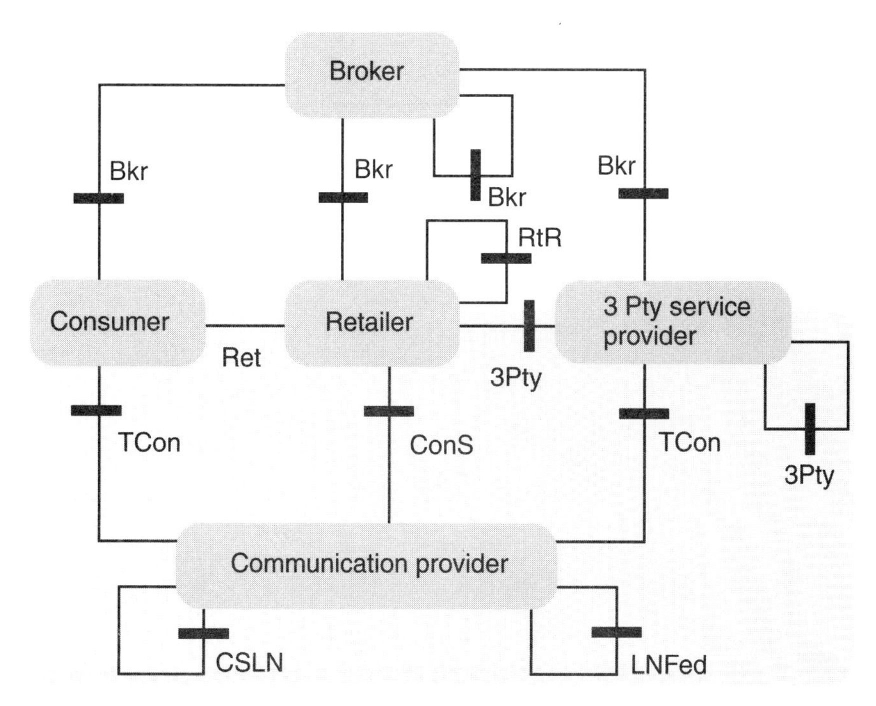

Figure 2.8 Scope of network resource architecture: roles

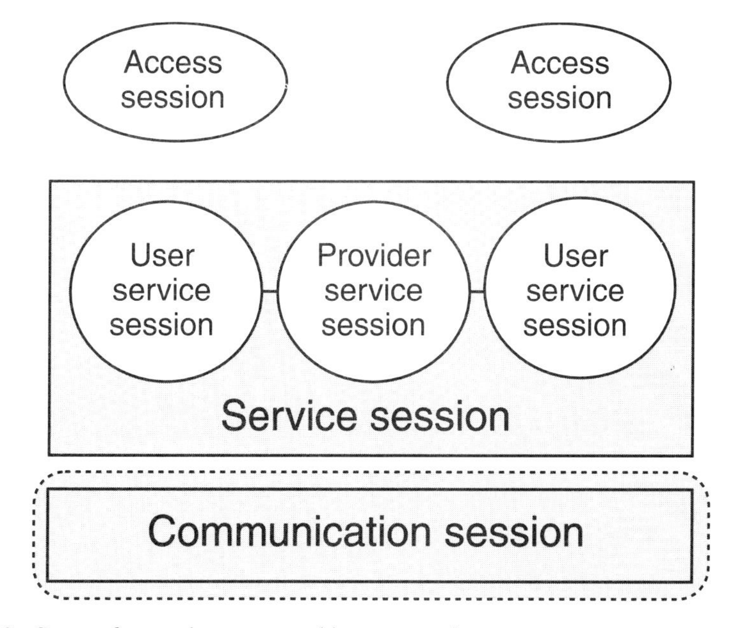

Figure 2.9 Scope of network resource architecture: sessions

high level of the service session, an abstraction of the connectivity is provided by the *connection graph*. This is an information specification using notions of vertices and ports (and lines to connect them) in different ways (see Figure 2.10).

- In the logical connection graph a vertex is a computational object, a port is a stream interface and a line is a stream connection
- In the physical connection graph, a vertex is a computing node, a port is a network access point and a line is a portion of stream in the network.

2.5.4.3 Connection management

The TINA network resource architecture defines how an identical service can be applied to different networks, using the appropriate connection management for each of them. This separation between the service and the network was organized to provide more flexibility

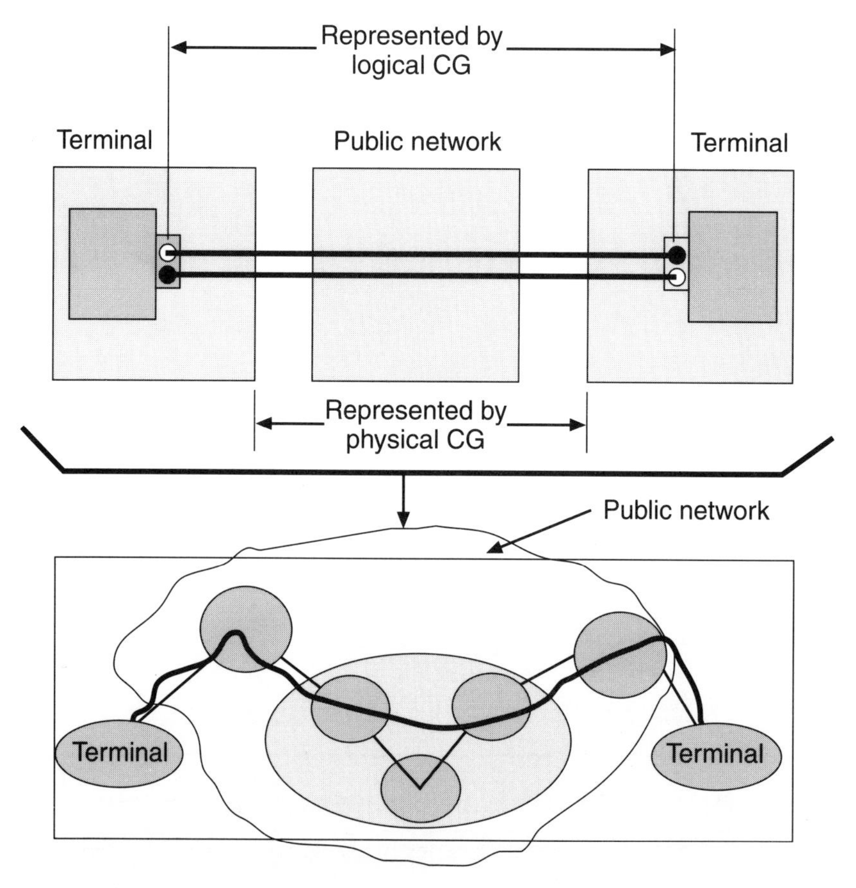

Figure 2.10 Physical and logical connection graphs

than the separation service to switching modeled in the intelligent network a few years ago. The separation interface is no longer a message-oriented call-dependent and trigger-fixed signaling interface; this time it is a logical representation of any physical link to any (call oriented or not) network. Of course, the connection management will not be able to provide the service management with rich command and information feedback on a network that has a very limited connection-performing interface. The connection management of TINA can therefore be considered a typically good level of information and actions for opening the control interface of a physical switch or of a cross-connect (see Figure 2.11).

The connection management components are (using top-down levels):

- Communication session level
- Connection co-ordinator level

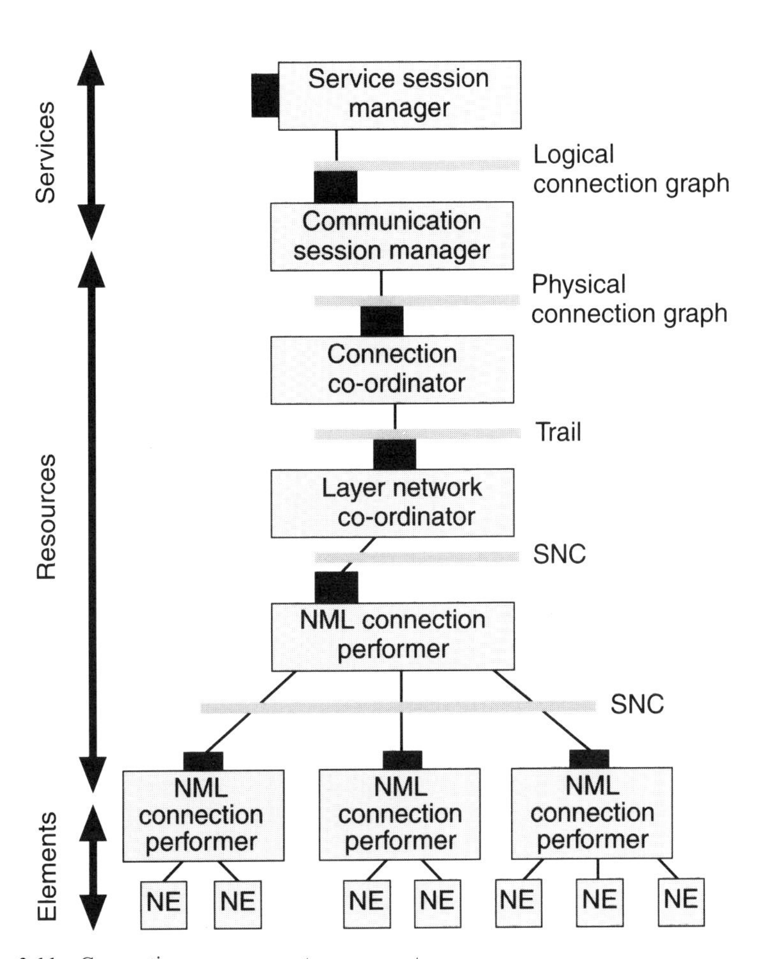

Figure 2.11 Connection management components

- Layer network level
- Network management layer and element management layer levels.

2.5.4.4 Assessment of network resource architecture

The NRA has been specified in detail (and has been the object of prototype implementation), but it is descriptive. An implementation, especially on existing networks, can use other principles.

Rules were given in 1997 on how to specialize it for serving connectionless networks like the Internet. The reference point ConS provides the interface between service and network, and was designed as the point of "clean-cut" between telecommunications manufacturers and the service factories, in association with Tcon for the terminal interactions. The NRA applies at this level.

2.6 GOING FURTHER

This section presents a list of references that are mandatory (e.g. TINA-C baselines) or useful for a more in-depth understanding of TINA. Many of these references are described in more detail in the following chapters.

2.6.1 TINA-C baselines

Architecture documents

- Service Architecture release 5.0 (end 97)
- Network Resource Architecture release 4.0 (end 97)
- DPE Architecture (1995)
- Naming and Addressing Framework release 0.2
- Management Architecture – see Service Architecture release 5.0
- Business Model and reference points release 4.0

Specification documents

- Service Component Specifications version 1.0
- Network Component Specifications version 1.0
- Network Resource Information Model version 3.0
- Ret reference point (1997)
- ConS reference point (1997)
- Tcon reference point (1997)

Modeling concepts

- Information modeling concept version 2.1

- Computational modeling concept version 3.2
- ODL Manual version 2.4.

2.6.2 Other TINA documents

- TINA Glossary version 2.0
- Guidelines into TINA version 1997
- Users guide.

All the previous documents are available in their public version at the TINA web site (www.tinac.com) or only for TINA Forum members in the TINA private server.

Chapter

3 Computing architecture

3.1 INTRODUCTION

3.1.1 Historical background

The TINA Consortium had among its goals to integrate into telecommunications the recent advances of computing technology, namely object orientation and distribution. A computer vendor could stop the chapter here and say "OK, the computing architecture is described in the Object Management Group as CORBA: you got it", an alternative being, just buy it with Microsoft OLE and DCOM. This chapter is a challenge: how to convince you that:

- First, telecommunications is a very special domain with such specific constraints that the general market computing product needs customization or considerable modifications before being used.
- Second, TINA brings gems to the software developers that are so precious that they should be included shortly to complement CORBA, and, in general, computing products fitting other domains like the avionics or the transport automation business.

To start with the TINA computing architecture, when the choice was made in 1992, most of the CORBA benefits like interoperability and a large library of object services were not available. It was not obvious that CORBA would have the lead quickly over other distribution solutions such as DCE. So the TINA choice to embrace CORBA was rather courageous.

For the next two years, TINA and the OMG worked in parallel: TINA reintroduced the OMG results as soon as they were stabilized. Reciprocally, the OMG telecom group examined the TINA proposals and issued a request for information and for proposals to get the TINA additions into the mainstream.

The TINA computing architecture is an expression of the requirements of the telecom industry *vis-a-vis* the information technology industry in order to obtain interworkable open platforms that will serve management as well as control platforms of any kind of telecommunications network elements. This chapter concludes with the benefits and the weaknesses of these requirements, and the chances to have them used partly or globally in the near future.

3.1.2 General

The reference model for open distributed processing (RM-ODP) is a frame for building distributed object-based systems with logical descriptions that provide flexibility for various implementations. In parallel with the RM-ODP model, TINA provides:

- At the design level, an information model and a computational model, and
- At the implementation level, an engineering model describing an abstract infrastructure for communication between TINA objects.

The main role of the computational model is to decompose each TINA application into a set of computational objects (COs) which are candidates for distribution. Some COs are presented in the chapters on service architecture and resource management architecture.

This chapter focuses on the engineering model of TINA, which maps the computational objects to the distribution infrastructure (DPE), through an abstract distribution model. The main benefits of this abstract distribution are to provide:

- A single communication space among applications
- A solution for heterogeneity and distribution

Communication with other objects is ensured whatever language and platform they are running on, and there is a conservation of the software programmed in the computing model, whatever object is distributed later on whatever physical resource: this is transparency.

Among the transparencies defined in RM-ODP, TINA requires from its DPE the access and location transparencies. The other transparencies are optional (see details in Figure 3.9). As the TINA DPE uses the Object Management Group CORBA technology as reference, the access and location transparencies are naturally provided by the object request broker (ORB).

As presented later, there are four main elements: the computational objects, the distributed processing environment, the kernel transport network (kTN) transporting information between DPE nodes, and the native computing and communication environment (NCCE) over which the runtime and the kTN terminations are implemented: see Figure 3.2.

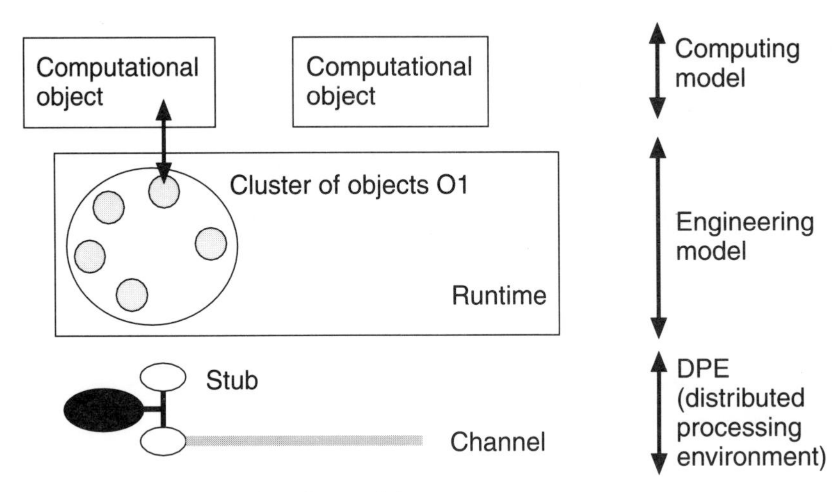

Figure 3.1 The TINA engineering model

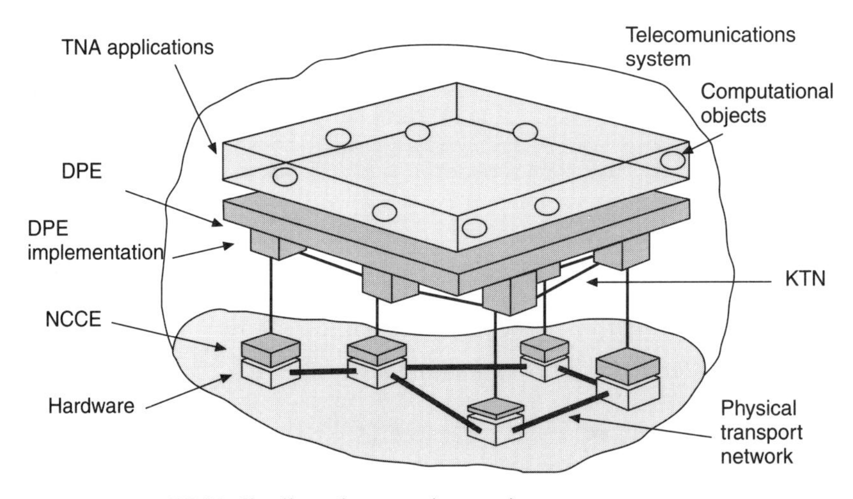

Figure 3.2 The TINA distributed processing environment

The DPE is detailed, with at first a study of its CORBA inheritance, and what can be done using CORBA only; in a second part we present the services and DPE-enriched model and behavior that justify equipping CORBA for telecommunications' special needs. The method and tools to design with the TINA engineering model are then presented, and we conclude with the issue of using the general computing market tools, as well as the standardization, for telecom specificities. But how far do you think that telecommunications can draw the blanket on its side?

3.2 DISTRIBUTED PROCESSING ENVIRONMENT: THE CORBA PART

3.2.1 DPE requirements served by using CORBA

The DPE model is an abstract distribution environment that serves many different requirements expressed by the applications from the different stakeholders specified in the TINA business model. The authors of the requirements are service designers, developers, and DPE system engineers.

The requirements can be classified as follows. First come general requirements on the DPE implementation, which are design portability and interoperability (Figure 3.3):

- Design portability means that the fundamental assumptions about the infrastructure will be supported by any TINA DPE implementation: thus the design can be ported from implementation 1 to implementation 2, without redesign.
- Interoperability means that TINA applications are compatible at both static and runtime levels, though they are running on different DPE instances.

These two requirements are met by using the TINA DPE infrastructure, whatever the profile of the platform as defined later. We could say that they are mandatory to claim minimal TINA compliance.

Second are explicit requirements as defined in the TINA computing model (these are the fundamental functional requirements assumed to be found on any computing environment). They are:

- Communication model as defined in section 3.2.3, including stubs and binding
- Access transparency to enable interworking across heterogeneous systems, and location transparency to permit several interface instances to vary transparently in position and space. The location transparency allows applications to access

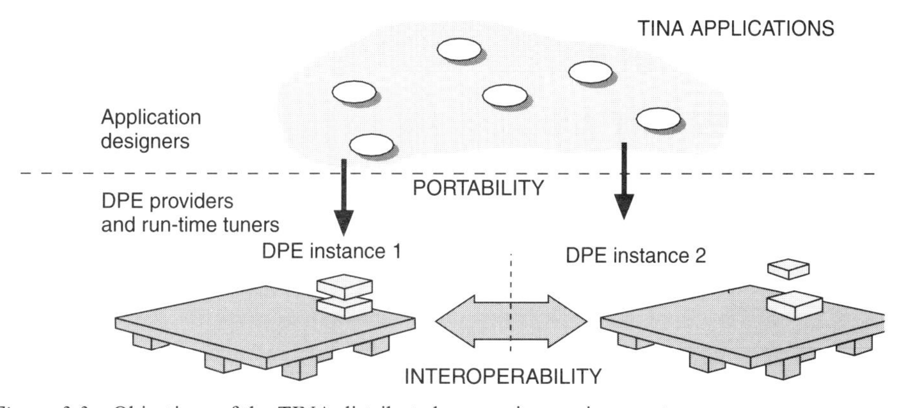

Figure 3.3 Objectives of the TINA distributed processing environment

services without using location information. One should not confuse this with increasing availability of service and load-balancing issues, which are tackled by replication and migration transparencies respectively. Both are not yet provided by CORBA, and remain optional in TINA.

Other requirements, functional or not, are detailed in section 3.3.1.

3.2.2 What is CORBA? A short presentation

The Common Object Request Broker Architecture (CORBA) is a standard defined by the OMG. The OMG was founded in 1989 and now encompasses more than 500 worldwide members. CORBA is one part of a larger architecture called object management architecture (OMA) described in Figure 3.4. In this architecture, applications involve objects, which can call each other to request a service through a software bus called the object request broker. The request can be passed to another ORB using the General Inter-ORB Protocol (GIOP) (see also section 3.3.5).

Applying CORBA to the DPE, TINA uses the ORB object services, and defines its own domain interfaces, as described in section 3.3. A key component of the standard is the Interface Definition Language (IDL); it enables us to define the interface independently of any language or compiler.

CORBA platforms offer common Object Services: Lifecycle, Events, Naming, Persistency, Transaction, Concurrency Control, Externalization, Security, Time, Properties, Query, Licensing. The CORBA Interworking Object Protocol (IOP) naming handles IORs: interoperable object references. The IOR numbering plan is administered by the OMG, like the URP numbering plan being administered for the Internet by the IETF. For further details on CORBA see Appendix 3.

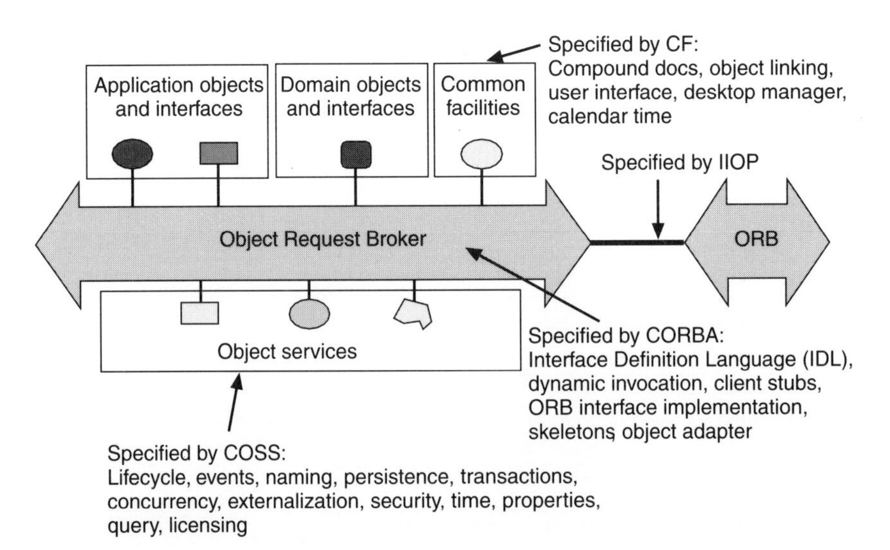

Figure 3.4 OMA (object management architecture) overview: what is CORBA?

The introduction of CORBA in the TINA DPE brings benefits in:

- Language independence: along with C++ and C, CORBA IDL supports bindings for SmallTalk, ADA95, and Java. CORBA acts as a language bridge, allowing the choice of language for a component without impacting the others, and keeping the global application integrable on-line.
- Openness and access to commonplace terminals: the generic standardized IOP interface enables interoperability with other non-CORBA domains, like Microsoft OLE, or SUN Java. Being able to port an application without modification to the source code, from workstations to PCs, Macintosh, or upcoming network computers, is a major benefit: it lowers costs and brings openness *vis-a-vis* the end-user hardware configuration.
- Design flexibility and reuse: it is possible to redivide components into subcomponents, to group objects in and out of the processes, without modification of the source code. Collocated objects behave like distributed remote objects, but reside in the same memory space, enabling request processing through direct function call. This allows a fine-grained adaptable modularity.

3.2.3 DPE kernel services: communication model

As explained in Chapter 2, the computational object interacts with other objects, either by invoking operations on an "operational interface" (blocking or non-blocking), or by exchanging stream flows on a "stream interface". CORBA only supports the operational interfaces. Streams can only be controlled as external black boxes, thus they cannot be considered as interfaces in the communication model.

To provide the operational interfaces, the DPE communication model involves four major elements: channel, binding, stubs, and protocol adapters. Using CORBA, only stubs and channels are available. CORBA provides only implicit binding without a binder. The OMG real-time group works on making the binding more flexible: this could happen to the TINA model later. As for the protocol adapter, the ORB protocol adaptation is done without individualizing the service, so that it could not be used to adapt for both operational and stream interfaces.

3.2.3.1 Channel

The interactions between computational objects which are part of different clusters have to satisfy access and location transparency requirements. The engineering mechanism which provides these transparencies is called a channel, following the RM-ODP terminology. The stub, the binder and the protocol adapter are the three fundamental functions needed to build the channel.

3.2.3.2 Stub

The stub supports the channel interfaces to the computational objects in Figure 3.5. There is not necessarily a one-to-one mapping of operations of the Cos, and of operations of the

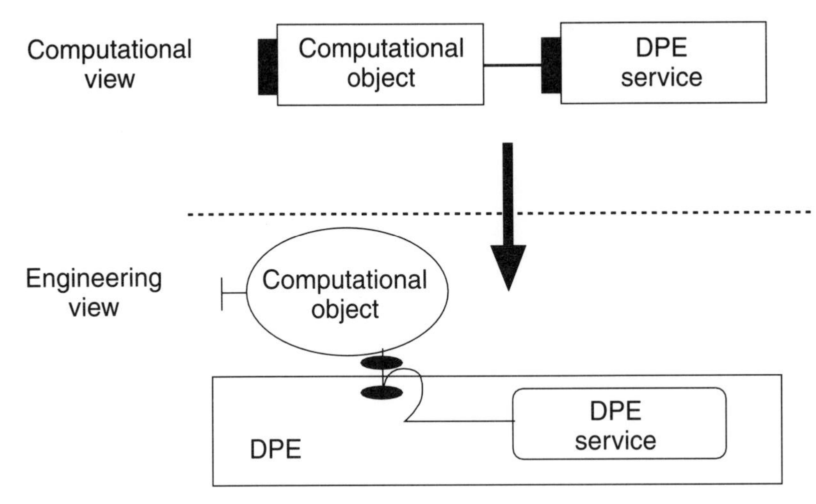

Figure 3.5 Representations of DPE services

interface supported by the stub (defined in IDL language). Alternatives are presented section 3.5, including the Microsoft DCOM with enhancements.

3.2.4 DPE services (object services)

The DPE services are defined generically, and are described independently of the native computing and communication environment that supports the DPE. They provide additional features to the basic DPE communication environment.

The TINA DPE services themselves can be described as computational objects: any object can invoke them through computational interfaces, as any other computational objects. Their computational and information specifications are part of the TINA specifications.

An application can make use of a limited subset of the DPE services. The set of required DPE services and other functional and non-functional requirements on the infrastructure for a given application corresponds to its "application profile". Not all TINA applications will use the same profile.

The choice of the infrastructure will be a trade-off between the different profiles of the applications. For instance, a trade-off must be made between the support of maximal distribution transparencies and the real-time performances requested in the non-functional requirements of the same application. It is unlikely that a platform supporting the full distribution transparencies set would be able to provide the same quality of service and performances as a platform providing only location transparency. Again it is expected to have platform profiles, defining the set of models, DPE services and quality of service that those platforms are able to support.

All the DPE services are not available in the OMG specifications because some of them serve specific telecom domain needs. Most are additions to the basic OMG specifi-

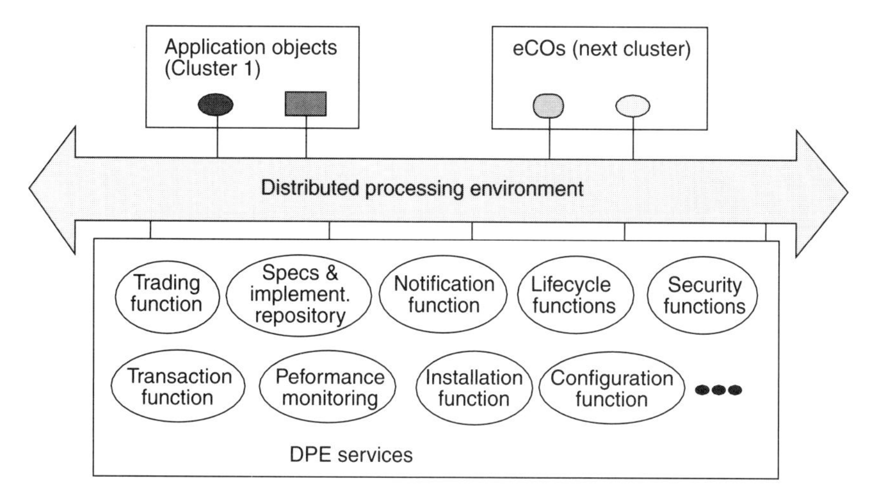

Figure 3.6 Set of DPE services

cations; for the most important ones, they have been proposed for addition to the object services by the Telecommunications Task Force of the OMG.

The set of DPE services is likely to change quite often: it is difficult to publish them, and this part of the book may be obsolete rather quickly (Figure 3.6). The following provides details of the DPE services accepted or available in the OMG Telecom Group at this stage. In addition, some ORB market products already offer proprietary solutions that serve part of the DPE services listed in section 3.3.2 as "non-CORBA", so the available DPE services list could be presented as larger.

1. Notification service: this enables objects to emit asynchronous messages without being aware of the set of recipient objects. Symmetrically, it enables an object to

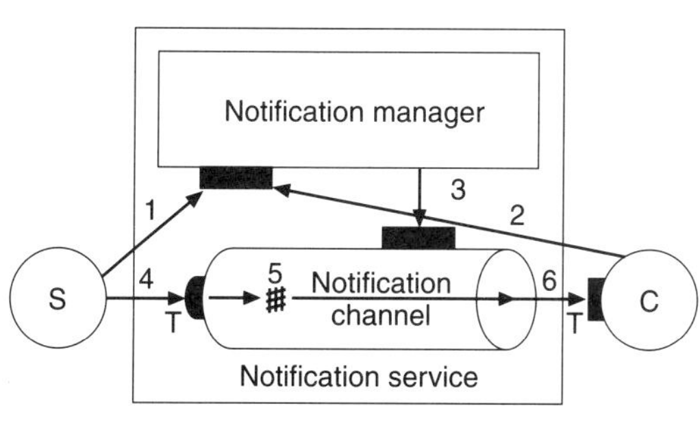

1. A supplier (S) wishes to emit a notification by invoking an interface of type T of the service. It subscribes to the service and gets a reference to a channel with the given interface.
2. A consumer (C) wishes to receive notifications of type T. It provides the service with a reference on an interface of this type and a filter (#) to the service
3. The filter and consumer reference are passed to the appropriate channel.
4. The supplier emits a notification by invoking the channel.
5. The notification is filtered according to the consumer constraints.
6. The notification passes the filter and the channel invokes the consumer interface for propagating the notification.

Figure 3.7 How the notification service works

receive these messages (notifications) without having a direct interaction with the emitting object. The notification can be seen as a broker between the notification emitter and receivers (see Figure 3.7). A typical usage of notification is the network management notification of alarms from a network element agent to the interested network managers. The notification service was proposed to the OMG Telecom Group as an extension of the OMG "Event Service", because the notion of notification is extensively used in telecommunications. Extension consists mainly of introducing notification filtering for reducing network traffic and addition of quality of service. The notification server receives requests to emit notifications from potential suppliers and provides as an answer a reference to a notification interface of the requested type. It also receives requests from consumers that express their intention to receive; these requests specify references, type of interface, possibly the identity of supplier objects, and constraints (such as filtering conditions). This is detailed in processes 1–6 of Figure 3.13. The notification service main operations are: consumer and supplier registration and deregistration, filtering and propagating notifications, filter update, managing quality of service (e.g. guaranteeing at least once delivery of notifications or ordering notification delivery according to priorities).

2. Trading services: trading is a service that provides a "matchmaking" function between a multicriteria request from a client and the various servers across the network which might offer the service required. Traders are useful in a large open environment where objects are dynamically added and retrieved on-line. Traders were very popular in the ANSA architecture. They are an attractive means to provide routing and scoping of interfaces all in one. Nevertheless, if the matching engine provides screening among a large population of servers, the performance of the trading may become an issue. The basic functions of this service are: Export service offer, Retrieve service offer, Modify service offer (service offers are characterized by properties and attributes describing the particularities of the offer). There is a late binding between importer and exporter (binding is not concluded at compilation time). The OMG trader provides a perfect match function and does not support an approximation (although it supports inheritance).
3. Security service: this server provides the logical operations needed to develop security in the applications, and thus in communication between objects. These security operations can be identification, authentication, credential handling (acquire, inquire, release), security context handling (initiate, accept, delete, token process), protection of dialog (sign, verify, seal, unseal). They could rely on the CORBA security and on the common secure interoperability services of the OMG, detailed here in section 3.2.8.
4. Transaction service: this server provides the operations needed to obtain an explicit or implicit transaction mechanism for updating data. A transaction applies to a set of actions on data and offers some of the ACID properties to it:

- Atomicity: actions are all performed successfully or are not performed at all (data is restored to its previous state)

- Consistency: a set of actions performed on a set of consistent data maintain the data in a consistent state.
- Isolation: other activities cannot see the modifications on data, as carried out by the transaction, until the transaction is completed.
- Durability: the modifications performed by the transaction will persist after it is successfully executed.

The transaction management operations are: identification, co-ordination, recovery, etc. A typical use of a transaction is to perform connection establishment atomically. Different transactional models can be handled (flat transaction closed-nested transactions, open-nested transactions, queuing transactions) allowing relaxation of some of the transaction properties for a lightweight service adapted to specific uses.

5. Log service. Reliability is a basic telecommunications requirement. Logging is the basic mechanism on which any reliability mechanism relies. Logging allows undoing or repeating actions, it is used for concurrency control, for recovery upon failure or for a distributed atomic update. At the time of writing, the OMG telecom task force has recently issued a request for a proposal on a logging service. This service should manage the writing of data with a logging semantics (i.e. keeping the causal arrival order of data, the reading semantics (reading data according to the associated causal order forward or backward), and log compaction when data kept become obsolete.

3.2.5 IDL

The Object Management Group has selected a notation called Interface Definition Language (IDL). As OMG focuses on support for object interaction only, this notation describes formally the object interfaces, but provides no support for the specification of object contents. The IDL syntax details the specification of the interfaces of entities supporting operation invocations; it includes interface type definition and structure and interface operations with attributes.

3.2.6 DPE stub generation

The IDL compilers make automatic generation of the stubs in various languages, beginning with C++. As communication stubs in each node and application represent an average 20% of the code of applications, using IDL with the automatic generation of stubs represents a very efficient way to decrease software development costs, and to shorten the time to market of services. Because these stubs are highly repetitive, and since multiple inheritance is allowed, using IDL is the way to ensure easy maintainable stubs in a C++ environment.

3.2.7 Telecom issue: CORBA used for real time

Having CORBA work in a real-time environment is not trivial. Some features consume a considerable time for execution: for example, marshalling, which provides a common format in the event of language or hardware/software heterogeneity along a remote access.

Also, the behavior of CORBA in real-time settings is insufficiently specified. To avoid these problems, one must pay close attention to elements that impact real-time performances, such as thread control and memory management. Additionally, the communication model underlying traditional ORBs lacks support for elaborate interactions (beyond the synchronous/asynchronous dichotomy).

Real-time telecom systems have specific needs that are not always adapted to the original CORBA model based on client–server interactions. Specifically, asynchronous calls are predominant, but with specifications different from those specified for CORBA's one-way invocations. Deferred synchronous invocations are also required, and CORBA offers them only through the impractical DII interface. Real time may also link with embedded systems, which add to the previous requirements a need for small footprints.

All the above requirements are being tackled in the Real-time Platform Special Interest Group's present discussions of the OMG. About half of the members of this interest group are from the telecommunications industry. There is a good chance that they will be offered solutions in the short term.

3.2.8 DPE security

No discussion of computational architecture would be complete without a look at DPE security and its implications for application and therefore service level security. There is an abundance of literature, both realistic and alarmist, on computer and telecommunications security which could be cited. There are also many examples of large fraud losses in telecommunications (notably mobile and premium rate services). Of the many aspects to security this section will consider only three: protection of subscribers from unauthorized service use (masquerade), protection from disclosure of service content (confidentiality), and protection of the provider from denial that services have been used (non-repudiation). Security is invariably an inconvenience for users, and therefore security designers should always restrain their more draconian temptations according to service ease of use, commercial value, and practicable fraud-detection limits.

In TINA access to and control of services is performed by DPE messages and therefore protection of these messages is fundamental to an overall secure architecture. The TINA business model assumes that kernel transport routes for messages might include untrustworthy network providers – in direct analogy to the Internet. Hence, irrespective of the invulnerability of each sub-network link there is still a need for end-to-end security that obscures from the sub-nets all information except that which is strictly necessary to route a message to its target end domain.

Other useful features are the ability to apply access control at a granularity of interfaces and operations and share by delegation established secure contexts among client and target side objects. Such DPE security services are clearly common to all the objects that wish to use them and this assigns the provision of object security to the DPE. In essence this means that client and target DPE nodes must implement interoperable DPE security covering compatibility details such as protocols, cipher algorithms and key lengths. Messages must be dispatched from the DPE into the kTN stack with protection of the target interface, operation and parameters together with any piggyback DPE service data,

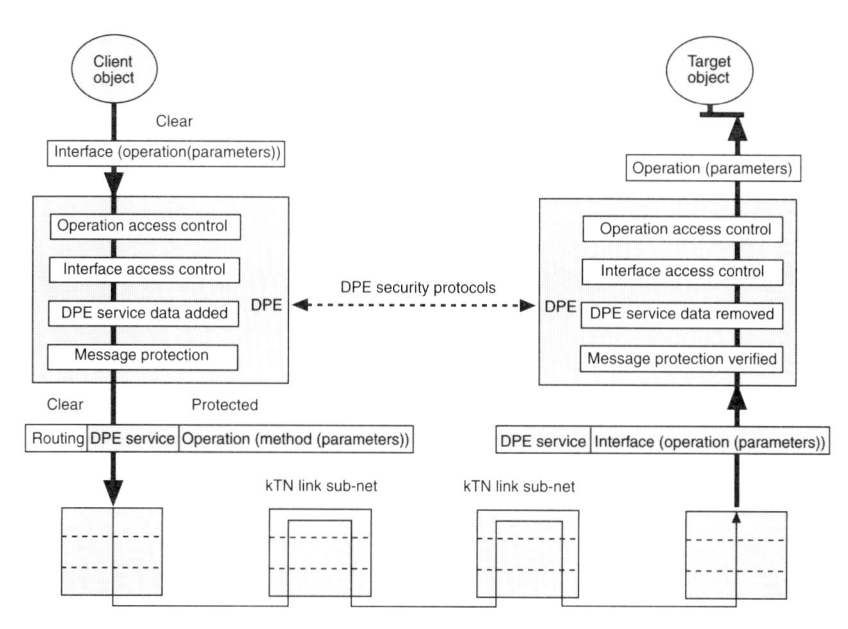

Figure 3.8 DPE security

leaving only routing data appropriate for the link layers to interpret. Obviously operation returns and out parameters must offer similar protection against the same threats (Figure 3.8).

All these splendid features ultimately depend on good authentication and secret or key distribution arrangements (whether within the TINA system or outside) and both trustworthy hardware and software at the end-points. These latter aspects may require particular technologies and hardware such as chip-cards or SIMM cards (used in GSM). Systemwide this results in various penalties including on-message latency, aspects of reliability (the system is more complicated) and larger DPE processes or security resources on each computing node. Nonetheless, it is still perfectly possible for TINA object designers to build-in additional security visible in interface and operation definitions, which might suit the particular enterprise security arrangements for the object services involved.

In conclusion, because of the fundamental and ubiquitous nature of DPE security TINA firmly devolves the common object security to DPE implementation and embraces security standards such as CORBA security services which comprise part of the common object services (see OMG CORBA Common Object Services, Chapter 15, Security Service, version 1.0). It follows that all TINA specifications expect these services to be present and therefore do not include security-related parameters in any interface and operation definitions.

At the time of writing no commercial ORBs support the entirety of these specifications. Additionally the lack of global encryption standards acceptable to all governments limits the levels of protection and interoperability on an international basis. However, the authors remain confident that both ORB technology will deliver and political differences can be solved with sufficient international breadth to achieve the vision.

3.3 ADDITIONS TO CORBA REQUIRED FOR TELECOM DPE

3.3.1 DPE requirements that are not served by CORBA

Only one of the mandatory applications requirement to the DPE is not served by CORBA (multiple interfaces) though, as mentioned above, some market ORBs have a proprietary solution to provide it. The deployment concepts and the support for object lifecycle management as described in section 3.3.6 are deemed fundamental in order to enable the deployment of objects and the management of the execution of objects on any TINA DPE.

Third and optional come the implicit requirements of the applications as usually needed in telecommunication distributed applications, which are to be selected upon application needs to profile the platform as mentioned in section 3.3.3. Optional are the DPE services also described in that section, such as the notification service and the transaction service. Optional also are the other ODP distribution transparencies apart from the access transparency, as mentioned in Figure 3.9.

Last come the non-functional requirements which describe the quality of service provided by the infrastructure. Among the quality of service requirements are:

- Performance for real-time usage
- Fault tolerance in the sense of fault avoidance
- Availability in the sense of permanence of service of a server viewed by its clients across the DPE
- Scalability of the architecture, in order to adapt to different network sizes or traffic usage, by presenting different architecture configurations

Provide an abstraction from the mechanisms essential to achieve communication between distributed objects, which aims at reducing the complexity of distributed programming.

Access:	Masks potentially different data representations between objects
Location:	Enables invocation of an object regardless of its current position or state
Federation:	Masks the differences of administration domains for interacting objects
Migration:	Masks the effects of the migration of application components
Transaction:	Masks the mechanisms for rolling back transactions
Replication:	Masks the replication of an application service from other replicas
Failure:	Masks the failure detection and recovery actions
Resource:	Masks the deactivation and reactivation of application components
Concurrency:	Co-ordinates concurrent interactions of several applications with one application component

Figure 3.9 ODP design transparencies

- Security, meaning the protection of assets and resources against potential risks.

None of these are mandatory. To claim TINA compliance, all of them, except fault tolerance and availability, are available, using CORBA. Fault tolerance and availability are also planned to have further specifications in the TINA architecture.

3.3.2 DPE kernel services: communication model

As explained in Chapter 2, the computational object interacts with other objects, either by invoking operations on an "operational interface" (blocking or non-blocking) or by exchanging stream flows on a "stream interface". To provide these interfaces, the DPE communication model involves four major elements: channel, binding, stubs, and protocol adapters. Besides CORBA, the TINA communication model adds explicit binding and protocol adapters (Figure 3.10).

3.3.2.1 Binding

Prior to any of such interactions *objects have to be bound*, regardless of whether computational objects are located or not on the same cluster or on the same node. In TINA, the binding can be either explicit or implicit. Implicit binding means that the application is not controlling or managing the bound. In explicit binding, the application explicitly requires the bound, and offers control and management facilities to the bound. This explicit binding involves a binding object.

The binder is in charge of maintaining the integrity of the binding, and the information about the channel such as the association context or the data buffer. The binder also validates the interface references, and can interact with a relocator when a binding error appears in order to update the interface location information. After an operation takes place, the binding disappears.

CORBA provides only implicit binding, without binders. Most of the ORB market products have an offer to control the binding from the application (explicit binding).

3.3.2.2 Protocol adapter

The protocol adapter supports a protocol that ensures that the interaction semantics are guaranteed for remote interaction between objects. It can be built above a connectionless kernel transport network (kTN) like UDP, or above a connection-oriented kTN like TCP/IP. It is an elegant way to present in a single model both operational and stream interfaces but very few implementation of this part of the model have yet been identified.

3.3.3 DPE services, TINA facilities, and DPE functions

3.3.3.1 DPE services (object services)

DPE services include but are not limited to the following:

- Lifecycle service

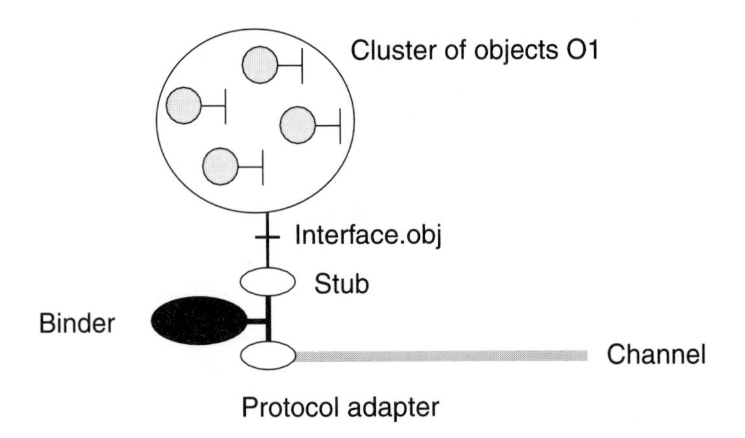

Figure 3.10 The DPE communication model

- Naming service
- Security service
- Notification service
- Performance-monitoring service
- Trader service
- Transaction service
- Concurrency control service
- Persistency service
- Query service
- Messaging service
- Licensing service
- Scheduling service
- Externalization service
- Property service
- Specification and implementation repository
- Installation service
- Configuration service

Some of these were detailed in section 3.2.3, as they are becoming part of the CORBA object services.

The following provides some details of other DPE Services, as shown in Figure 3.6, which are not likely to become OMG accepted in the near future. Of course, this information may become obsolete and the reader can check whether some of them have become OMG telecom standard object services.

- Specification and implementation repository: this is a place to store the specifications of an object, its interfaces, operations, data types, etc. in a defined release. It serves the object lifecycle management, and can be used, for instance, by a factory to create instances. The basic functions of this service are: add, remove, update and retrieve for interface templates, for object templates, for relationships, and for implementations. It also provides a lookup contents function, using the templates' identifiers. The naming scheme for the objects, as mentioned above, is Full Distinguished Names (FDN) of telecommunications network management.
- Lifecycle functions are detailed in section 3.3.6, as they support the TINA objects lifecycle.
- Performance-monitoring service: performance monitoring is the service able to deliver to applications information on the performance of the infrastructure. It requires from the infrastructure "call-backs" that transmit information and management orders to and from the specific network resource acting with a given performance. The performance-monitoring main operations to the applications are: resource workload and usage indication, resource allocation delay, resource rejection rates indication, resource failure rate indication, communication throughput and response time indication, transmission throughput and error rates indication. The performance-monitoring operations towards the infrastructure are: activate, schedule or suspend performance-monitoring activities, retrieve performance information, specify performance-monitoring rules and criteria, adjust the network resource to improve performances.
- Installation service: this service is used for the deployment of engineering Computation Objects (eCOs), locally or remotely from the object father of the creation.
- Configuration service: this service is used for providing information on the location, cluster participation, and activity status of the TINA objects.

3.3.3.2 DPE facilities (common facilities)

DPE facilities are generic computational objects which are offered in addition to the DPE. They are higher-order and more complex services than those previously mentioned. Nevertheless, some DPE services that have an application flavour can become DPE facilities in another classification, such as the installation service and configuration service. Typical examples consist of a generic user agent, or a human–machine interface server. They could also provide topology, or quality of service management facilities. TINA DPE facilities are not prescriptive: they are optional.

DPE functions include basic functions like the DPE management functions and a co-ordination function that co-ordinates resource usage. Basically they deal with platform software management.

3.3.4 Naming and addressing

The different elements of the TINA engineering model need to be unambiguously named with globally resolved names. Naming is needed for:

- Nodes
- Capsules
- Clusters
- eCOs
- Object instances
- Object interfaces

The approach taken is to try to adapt the existing frameworks: CORBA IOR for the objects, and possibly Full Distinguished Names from the telecom network management for larger entities.

3.3.5 Kernel transport network (kTN)

The kernel transport network is the underlying transport network that interconnects the DPE instances. It is to the DPE what the GIOP (General Inter-ORB Protocol) is to the ORB (Figure 3.11). No assumption is made whether a kTN is connectionless or connection-oriented. The quality of service requirements are expressed by the applications to the kTN. Since the kTN is equivalent to the telecom signalling links, the level of reliability of service should be similar.

In order to provide information transfer, the DPE might need to establish connections through connection-performing objects. Of course, they cannot rely on the main con-

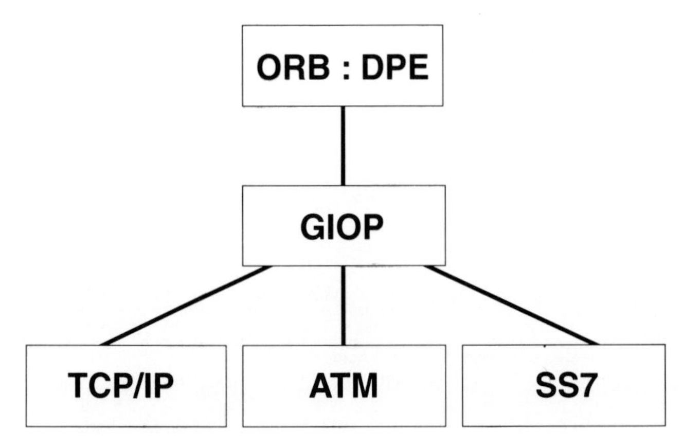

Figure 3.11 The kTN structure

nection performer: that needs the DPE itself to convey connection orders (chicken and egg initialization). "Bootstrap" procedures are needed for kTN interoperability.

The kTN will be based on GIOP (General Inter-ORB Protocol, from OMG). GIOP's most common mapping is presently to TCP/IP (IIOP) over LANs or ATM, and we expect a similar mapping to Signalling System 7 (SS7) for public telecommunications networks and to ATM. Further projects elaborate on IOP over ISDN and GSM.

The kernel transport network (object interactions for control and maintenance) is logically separated from the transport network (main network connections, streams). This enables a separate engineering of both sets of paths and may be required for regulation reasons to differentiate the kTN operator.

It is possible to imagine that the kTN is offered by a separate operator, different from the service or from the connectivity provider. Therefore, the notion of a kTN reference point (also called a DPE reference point) is important. It is detailed in Chapter 6.

In order to maintain the flexibility to offer the kTN to another operator one should make early plans for kTN charging! Accounting for the kTN is also needed for planning and traffic history purpose. The provision for accounting data can be made within routers for TCP/IP-based kTN, or within signalling management points, for SS7-based kTN.

TINA DPE (including CORBA) over SS7 can be considered as a smooth migration of a message-oriented signalling towards an object oriented signalling. If the SS7 signalling point is not built to support the DPE/CORBA infrastructure, it can handle the SS7 lower layer and pass over the DPE syntax to an interworking unit. The interoperability with other software distribution technologies like OSF DCE, Microsoft DCOM, Java, etc. is contained within the scope of the evolution of the TINA DPE architecture.

3.3.6 Object lifecycle and deployment concepts

3.3.6.1 Object lifecycle

The object lifecycle defines the concepts and interfaces for supporting the management of a TINA object: deployment, creation, deletion, activation, deactivation, moving, and copying. It is larger than the OMG lifecycle since OMG's service is only covering an instance lifecycle (see Figure 3.12). The object lifecycle operations have to be offered on each DP node, and are essential for compliance and interoperability. They must be offered to local TINA applications, as well as to remote applications. Several stages have been identified in the lifecycle of an object (see Figure 3.12).

One could describe a generic object lifecycle as being composed of the following steps:

1. The developer designs the TINA application as a set of COs, described in the ODL language (see section 3.4) for the interfaces, as well as for the object content, and uses an ODL compiler that will precompile in IDL to generate stubs.
2. A computing object code is written and compiled to form executable codes.
3. The executable codes are downloaded in the appropriate DPE nodes by the installation service (deployment phase).

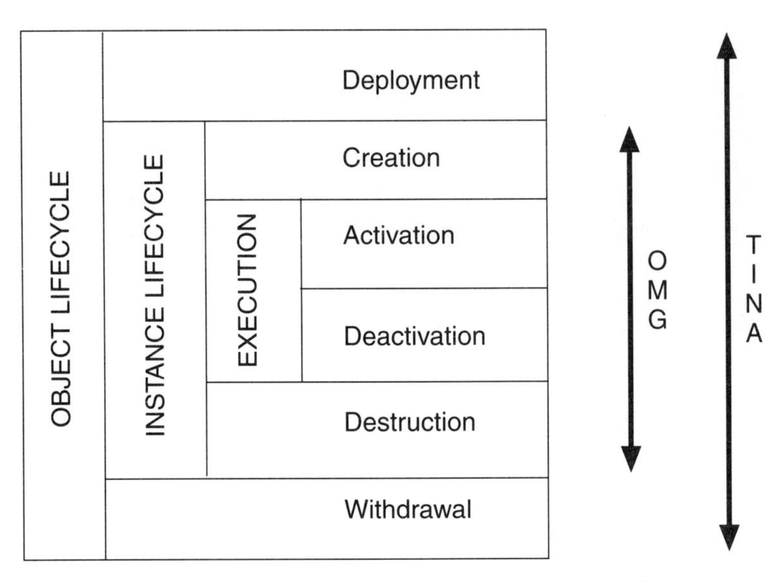

Figure 3.12 Stages of the object lifecycle

4. The clusters = sets of eCO instances are created by the object factories (creation phase).
5. The clusters are activated: they can run for service execution (activation phase).
6.–8. For the end of the lifecycle, the clusters are deactivated, then the cluster can be destroyed, then the eCOs can be withdrawn.

Reservation of object resources before activation has not been considered useful.

The deployment phase makes use of an installation service (or installer) offered by the DPE. The installer is able to select the eCO or the cluster of eCOs (specified in the request to the installer by their object name or cluster name), and to deploy them on nodes, which have been identified in the request. As a result, the instances will receive an identifier and be deployed on the requested DPE nodes. The installer main operations that the father applications may invoke are: object deployment, cluster deployment, object withdrawal, cluster withdrawal ... All these operations take as input identifiers, which are compliant with the TINA object and interface naming and numbering plan (see next section).

The creation phase makes use of object factories. Unlike the installer, object factories are not DPE services, they are normal TINA objects.

From a computational viewpoint, the creation of an object is performed by instantiating an object template that provides the information relevant from the computational viewpoint. Further information may be added in an "engineering" template for the same object creation.

The object factory is specialized for certain families of objects. For instance, there will be a service factory in the service management part, another in the communication session management part, and so on. They all rely on a DPE generic factory and this can be the CORBA generic factory. The object factory main operations are: create object instance (upon presentation of an object template, with required specialization) locally or remotely, destroy object instance locally or remotely, create cluster instance, destroy cluster instance, add object within a cluster, remove object from a cluster.

Object activation is done by an object internal activation thread or by external invocation of one of its operations. The active object management is performed by the object configuration service, which is one of the DPE services.

Object configuration provides object configuration static information such as object node identifier, object capsule identifier, object cluster identifier, object interface references list, object service attribute values list. It also provides configuration dynamical information such as object binding status (list of interface references to which it is bound) and object status (activated or not).

3.3.6.2 TINA applications and deployment concepts

As mentioned above, a TINA application is composed of a collection of computational objects (COs). These computational objects are represented in the engineering model by eCOs; an eCO is an engineering CO that corresponds one to one to a computational CO).

TINA Application Services
- Application developed according to TINA-C architecture
- Example: CSCW

TINA Facilities
- Generic TINA Application Services
- Example: user agent

DPE Services
- Services provided by some DPE platforms as a support to TINA services
- Example: Trader

DPE functions
- Example: DPE management

Native Communication and Computing Environment

Kernel Transport Network
- Provides interconnection of DPE

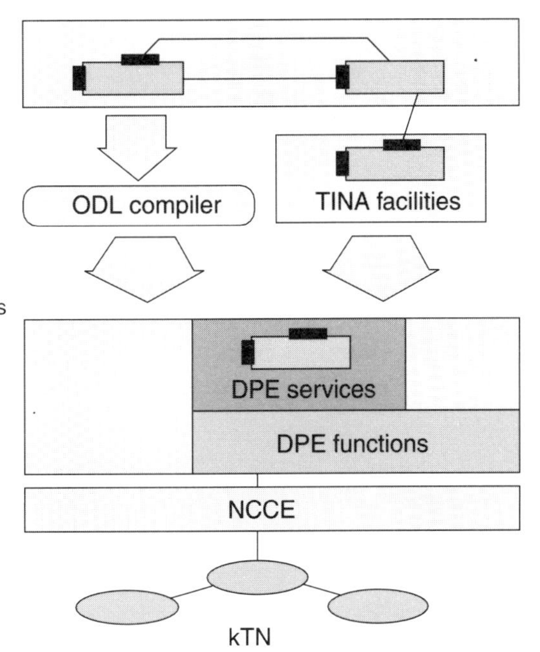

Figure 3.13 Deployment concepts

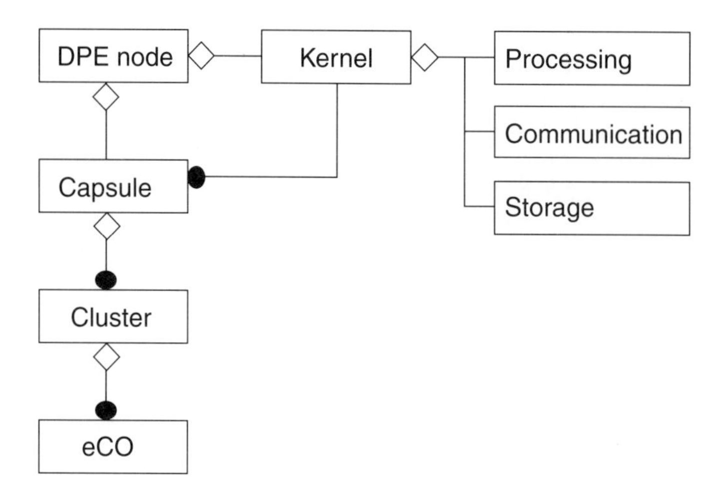

Figure 3.14 TINA global engineering architecture

Unlike a computational object, an eCO is able to interact with other engineering objects which do not come from the application, such as traders.

When deploying the engineering object instances, engineering objects are grouped in clusters, which are grouped in capsules included in the DPE nodes (Figure 3.13).

DPE nodes are, for example, single-processor systems of distributed systems. Capsules typically are heavyweight UNIX processes or a virtual machine modeled, for instance, by actors (chorus operating system). It is the engineering unit of resource allocation, not of distribution, as stated in RM-ODP.

Clusters consist of a collection of objects (it could be an object group if they also share instantiation commonalities). A cluster is the unit for activation, placement and migration, but not necessarily the unit of instantiation. This fits with the RM-ODP cluster notion.

3.3.7 DPE tools and method

The TINA developers are provided with a support for execution (Figure 3.14) which consists of:

- A telecom domain-specific architecture standard: the service architecture, and network resource architecture
- A common object model: the computational model
- A common specifications language: the ODL
- A common behavior description language (not mature yet)
- DPE server functionality interfaces and semantics
- A common kernel reference model: the engineering model

3.3.8 ODL

Because TINA deals with telecommunications applications and not only with object interaction (OMG focus), it was needed to provide developers with a specification language that would serve not only the object interfaces (IDL) but also the object internal logic. If this larger-scope language could be used to generate automatically the object code beside the communication stub code, such an economy would even be much more attractive than the 20% mentioned earlier when using IDL. Unfortunately, this approach was too ambitious, and not compatible with the Consortium's timeframe, so ODL was finally targeted to IDL additions needed to match with the TINA object model.

Possible additions for an abstract expression of the real-time constraints, of the configuration and fault-tolerance principles, and for data management may be discussed later; the basic idea will be kept to maximize the code's portability by extending the abstract specification of the applications within what is mature in the state of the art. The global abstract expression of all the object logic and behavior is still an open domain for academic research. As an example, ITU-T has started work on the integration of SDL'92 with IDL or ODL.

So, basically, here the TINA language is a superset of IDL, enriched with the functions that were missing in order to cope with the TINA object model (slightly different from the OMG object model): object groups, multiple interfaces per object, quality of service requirements, and streams, supporting telecommunications applications. This concrete syntax, supporting the computational specifications of object templates, interfaces templates, and object group templates, has been developed and is called *TINA-Object Definition Language* [TC-MO 96]. TINA-ODL is used in:

- *Application specification and specification re-use* (at development time): when developing computational specifications, an application developer needs to be able to describe a TINA application in terms of computational modeling concepts (object groups, objects, operational and stream interfaces, and data types). In addition, development effort can be reduced if existing computational specifications of TINA applications can be re-used in the specification of new applications.
- *CASE tool development*: Application specification, and the application development process, can benefit from the type of automation typically offered by CASE tools. In order to begin constructing such CASE tools, the language of specification needs to be available and supportive.
- *Application execution and interaction* (at run-time): In order to support dynamic binding and dynamic configuration management of systems, a common syntax is needed to describe the entities involved. The availability of a common language to describe operational and stream interfaces enables the definition of compatibility for interfaces, which is important for binding purposes. Interface type management needs to be extended to object types as supervisory and control systems for distributed applications become more sophisticated. Support for application execution requires the definition of run-time interaction constraints, which are naturally specified in a language like TINA-ODL.

3.3.8.1 Objects syntax

Freedom is offered to the developer of computational specifications using TINA-ODL for independent declaration of interfaces, objects, and object groups. Each interface template in TINA-ODL may be self-contained, and may be re-used in any number of object templates. Similarly, object templates may be specified as individual units, and re-used in any number of object group templates.

An object template specification comprises a *header* and a *body*. The header supports the declaration of the object identifier. TINA-ODL allows for the specifications of object template inheritance within a template header, to support specification re-use and to provide a mechanism for defining compatibility via sub-typing relationships. The object template body encompasses the following sub-parts:

- *Object behavior specification*, in the form of a natural language. It describes the role of an object in providing services via each of its interfaces.
- *Object quality of service specification*, in the form of typed parameters representing quality of service attributes. The parameters are intended to allow the specification of the "level of service" supported by a particular object instance. Such information may be defined dynamically by the management system responsible for initiating object creation. The values of these parameters may be negotiated when an object is instantiated or changed when the parameters are altered during the lifetime of the object. Examples of such parameters are security constraints on the object location or reliability constraints in the form of specification of the maximum admitted probability of failure.
- *Object initialization specification*, in the form of the name of an interface template, among those supported by the object, which may be used for initialization. A reference to this interface will be returned to the instantiator of the object template.
- *Required interface specification*, in the form of a list of interface templates. It specifies interface types, which will be used by the object to provide its services.
- *Supported interface specification*, in the form of a list of interface templates or names of interface templates. Instances of interface types declared as supported may be offered by instances of objects being defined.

3.3.8.1.1 Object group syntax

An object group template specification comprises a header and a body. The header supports the declaration of the group's identifier and possible inheritance relationships. The object group template body encompasses the following sub-parts:

- *Group behavior, quality of service*, and *initialization specification*: The behavior and quality of service specifications of an object group strongly parallel that of an object. The initialization specification of an object group is slightly different from the object one in that it specifies an object template to be instantiated instead of an interface template.

- *Group contract specification*, in the form of a list of interface templates. Contract interfaces are the interfaces that are visible to entities outside the object group.
- *Supported objects and object groups specification*, which describes the object and object group templates that can be instantiated within the object group.

3.3.8.2 Interfaces syntax

The specification of an interface template, which can be an operational interface type or a stream interface type, comprises a header and a body. The interface header supports the declaration of an interface identifier. As in object template specifications, TINA-ODL allows for the specifications of interface template inheritance relationships. The interface template body encompasses the following sub-parts:

- *Interface behavior and usage specification*, in the form of a natural language text. It describes the behavior of the interface and the constraints on the use of its operations or streams.
- *Interface quality of service specification*, in the form of a typed parameter representing quality of service attributes. The type of the parameter is intended to allow the specification of the "level of service" required by a particular interface instance. The value of this parameter may be set or negotiated when an object is instantiated, or may be altered during the lifetime of the object.
- *Interface trading attribute specification*, in the form of a typed parameter representing trading attributes. These attributes describe the properties of an interface as used in constraint specifications when trading for interface references. For each interface type (an operational or a stream interface) it is possible to specify parameterized qualities. These qualities are typically specified by a server (or another object acting on its behalf) and used when exporting an interface reference to a trader. A client of that interface type may express its requirements to the trader in terms of this set of parameters.
- As appropriate, an *operational interface signature*, or a *stream interface signature*.

3.3.8.2.1 Operational interface specifications

An operational interface signature comprises a set of interrogation and announcement signatures, one for each operation type in the interface. The information specified is identical to the one specified when using IDL.

3.3.8.2.2 Stream interface template

A stream interface signature comprises a set of flow types called *action templates*. Each action template for a flow contains the name of the flow, the information type of the flow, and an indication of whether it is a producer or a consumer (but not both) with respect to the object which provides the service defined by the template. A stream interface signature specifies the following information (defining the stream server from the viewpoint of the client):

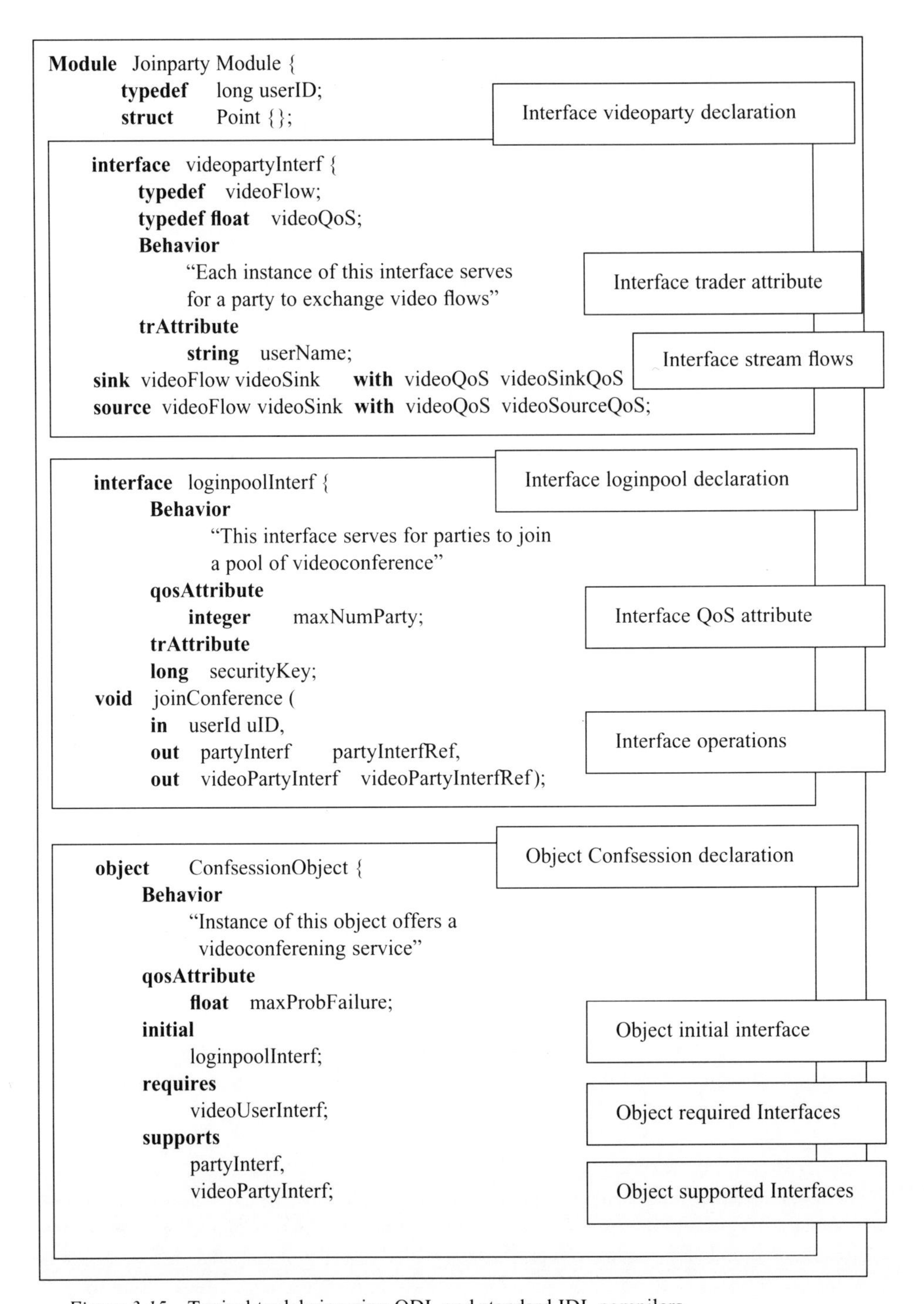

Figure 3.15 Typical toolchain using ODL and standard IDL compilers

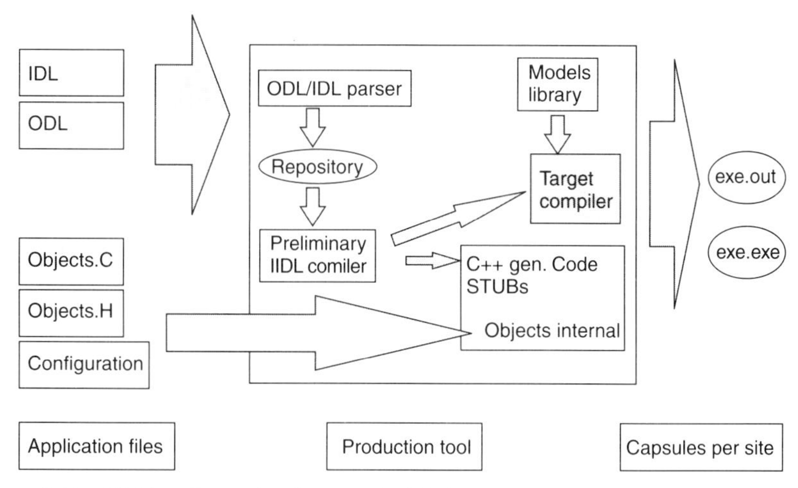

Figure 3.16 TINA software development toolchain

- *Signature of each stream flow* that can occur at instances of the interface type. It specifies the frame structure and coding details associated with the flow, and a producer/consumer attribute that specifies whether an object that offers the interface is a producer ("source") or consumer ("sink") of the stream flow.
- Name and type of each *service attribute* associated with each stream flow.

3.3.8.3 Brief ODL example

The following example of the ODL syntax describes an operational interface that enables a telecommunications user to join a pool of participants in a multimedia conference. To simplify the example, only the video part is declared in Figure 3.15. A typical toolchain for the development cycle using ODL and standard IDL compilers appears in Figure 3.16.

In summary, we have successive mappings between three levels of objects:

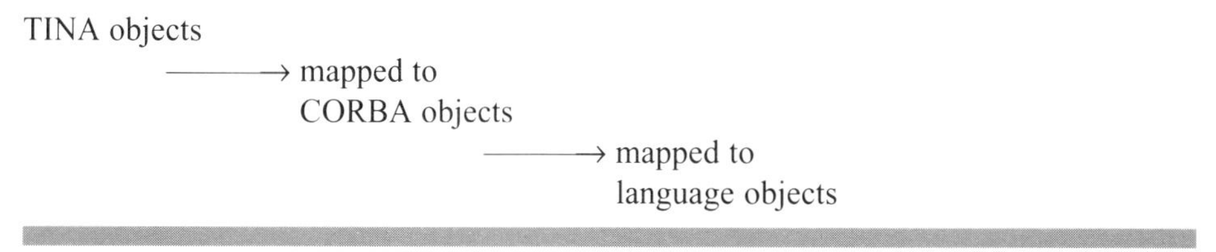

3.4 CHOICES

What is a minimal DPE that is TINA-compliant?

3.4.1 Choices: plain CORBA, TINA DPE or other solutions

The main differences between CORBA and the TINA DPE were detailed above. Briefly, TINA adds to CORBA the object lifecycle, the DPE services which are not common object services under discussion, i.e.

- Specification repository
- Performance monitoring
- Configuration service

some lightweight profile of the COS services (e.g. a lightweight persistence service) or some extensions (e.g. transaction model for transaction services), and a few architectural attractions that the DPE proposes in the object model:

- Object grouping
- Multiple interfaces per object
- Streams integrated modeling
- Dynamic and late binding
- Quality of service expression
- Real-time computing expression over CORBA

Fortunately, most of the ORB commercial products offer rather more than the OMG standard, for instance in lifecycle and in bindings, but for portability and maintenance reasons it may be wise to keep to the only parts that are standard implementation.

This means that TINA is a strong lobby to get ORB products to evolve to major telecom needs. Nevertheless, until these telecom features are standardized, TINA industrial products will most probably somehow lag behind TINA, using only what is tooled by the general object orientation market: stream integration and multiple interfaces may stay in the refrigerator for a while.

Meanwhile, the required telecom features which are easily solved at the application level like TINA object factories, installation servers and object configuration service will easily be deployed in the early TINA products.

3.4.2 Plain IDL, ODL or other solutions

It should be stressed that OMG-IDL and other IDL languages do not fully support the concepts defined in the TINA computational model. In particular, while OMG-IDL shares most of the objectives of TINA-ODL (support to application specification, application re-use, and application interaction), OMG-IDL does not provide any support for application execution. A number of TINA computational modeling concepts supported by TINA-ODL do not find any equivalence in the OMG object model and hence OMG-IDL. TINA-ODL extends OMG-IDL in the following ways:

- At the object level, TINA-ODL offers support for the QoS requirements, objects with multiple interfaces.
- At the *interface level*, TINA-ODL offers support for the definition of stream interfaces, interface behavior, and operation and stream QoS requirements.
- At the *operation* and *stream levels*, TINA-ODL offers support for the definition of stream signatures.

- At the *application structure level*, TINA-ODL supports aggregation of larger abstractions, called object groups.

TINA-ODL is a superset of OMG-IDL: the syntax defined in TINA-ODL for operational interface declaration encompasses and supports all the rules defined for OMG-IDL. This implies that the OMG-IDL specifications are part of TINA-ODL specifications (as operational interface declarations).

We should also note the different abstraction level between OMG (pragmatic and interface oriented) and TINA that provides a lifecycle-organized object model. As an example, there is a difference of granularity that exists in a TINA object and is unknown in OMG: the interface is the granularity for interaction, usage control and expression of the communication model as in OMG. But the TINA object is the granularity for object re-use, and for configuration management. In that sense, CORBA objects can be seen as TINA interfaces, with IDL as a subset of ODL, but in this case higher-level management tools (like ODL to IDL compilers) are required.

Using a superset from IDL is not a difficult process as long as the ODL parser and extraction is a simple tool. It makes it worth using an ODL as long as the additions are really saving time in re-use, portability and interoperability. A large part of TINA's early deployment experienced ODL with very simple parsers (a few thousand lines of code for the tool) and found benefits in it. But the ODL used has been diverging from one vendor to another. Some stress the real-time characteristics abstract expressions, others focus on quality of service abstract expression and others on stream abstraction.

This is why the discussion for standardization of the telecom ODL is a rich and open issue. The debate was stressed on the first TINA DPE implementation using CORBA. Nevertheless, any large implementation of DPE will be based on a variety of implementation technologies and will include heterogeneous ORBs, such as a combination of CORBA, DCOM and DCE.

3.5 COMPUTING ARCHITECTURE STATUS AND STANDARDIZATION

3.5.1 What is stable?

Briefly, one could say that what is accepted by the OMG is stable. That is:

- A first level of industrial stable and tooled status is based on the standard CORBA DPE, adding those TINA DPE servers which have been well specified and discussed in OMG, like notification server, transaction server, and security server; the two latest servers should be taken in the OMG specification status. Besides this standard kernel, one should add home-made (even limited) factories, repository, installation and lifecycle servers; object groups can be used with limited effort (there are attractions in most CORBA products for it), as they are really useful for making a sturdy and clean design of large telecom applications.
- A second level of stable standard specifications with expected but still unavailable industrial tools encompasses the usage of objects with multiple interfaces

and streams, an expression of real-time over the ORB, dynamic binding, a logging and possibly ODL. Configuration servers are also expected to be discussed soon in OMG, for the persistency and permanence of service parts.

- What can be classified as unstable or not yet completed specifications are IDL complex types, interfaces exception handling (currently home-made), network element callback handling (TMN usage), accurate syntax for expressing quality of service, performance management, and functions for DPE node management.

3.5.2 What should be completed in the near future?

The API for the DPE usage by application is not yet complete. It should be refined in the future, especially for exception handling, quality of service, and for the DPE servers which have not yet been specified.

The scalability of the DPE for small terminals has not been completed. The option taken by projects involving small-end customer premises equipment has been to stop the DPE before the terminal, either in the network server or in a network termination at the customer premises, or near the last operation switch. If this is to be a major TINA implementation option, then the mapping from TINA to the non-TINA terminal (Java or equivalent), should be given, including the mapping over the user network interface part of the signalling, if any.

This issue is also closely related to the definition of the kTN and the availability of GIOP engines. It is foreseen to obtain such engines with a minimal footprint that would not involve a complete ORB. This will make very small terminals, like mobile handy phone sets, able to "talk" to a TINA DPE. A definition of the capabilities of such a "TINA terminal" is needed.

Migration from the legacy telecommunications network to TINA servers or TINA end-to-end services involves mapping the DPE (and the TINA reference points over it) to the different major telecommunication interfaces: INAP for the intelligent network, CMIP for the telecommunications management network ITU-T compliant, SS7 network signalling, ISDN-UNI and B-UNI, etc. (Figure 3.17).

The ORB part of these mappings has already been discussed in various fora: CMIP/CORBA in X-Open Joint Inter Domain Management task force, INAP/CORBA in OMG. Completion is expected in the short term.

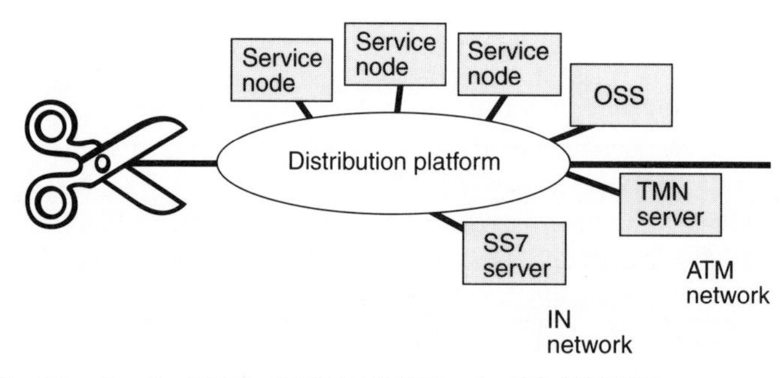

Figure 3.17 Mapping the DPE to IN/SS7+INAP and to TMN/CMIP

3.5.3 DPE benchmarks

There have been several benchmarks comparing:

- The different OMG-compliant ORBs, with
- The Microsoft DCOM, and with
- The new real-time profiled ORBs for embedded usage.

Also, measures are available (see the ACTS project ReTINA) for:

- A TINA DPE based on an ORB, used with a TINA-compliant object model including streams, and enriched with notifications, initialization and repository servers
- Using this DPE with the TINA connection management and service architecture.

There has to be a deal between the level of functionalities of the DPE, which spare development time, and the level of real-time performances, which may require a limitation in the number of OMG CORBA services (COSS). The first services with the full TINA architecture have shown appropriate results for services like videoconference with dynamic sessions and link establishments (see the ACTS VITAL project, in Chapter 8).

3.5.4 OMG standardization of TINA

In order to see rapid commercial availability of TINA-compliant DPEs there should be an extension of commercial CORBA platforms. This would be facilitated by the integration of TINA computing architecture concepts into the OMG standard. Contributions on behalf of TINA-C have already been injected into this group, and recognition of some of its requirements has already been achieved. It is of paramount importance that the TINA-C and the OMG communities join efforts to achieve complete support for requirements and concepts underlying the telecommunications information networking architecture.

3.5.5 Probable DPE parts to become standard

As mentioned above, it is very likely that the following specifications will become OMG-Telecom Group standards:

- The TINA object model (streams, object group, multiple interfaces per object, quality of service expression)
- Part of the TINA ODL

These arise from two related activities of the OMG:

- Control and management of audio and video streams: this provides concepts of streams

- Multiple interfaces; this provides support for the aggregation of interfaces (CORBA objects), and encompasses part of ODL for the specification of these additions.

3.5.6 Competing standards beside OMG

Work is ongoing in the ITU-T, study group 15, for the application of ODP principles to the specification of telecommunications management systems [ATMF95]. As a support to these specifications, an ODL-like language is in the process of being defined. This language does not currently provide support to all the concepts of TINA-ODL, such as the concept of streams and object groups. The ODL is now discussed as part of the ITU-T SG10 Question 2. But under the same denomination of Object Definition Language, the content may largely vary.

Also mentioned was the Microsoft approach with DCOM, a similar object distribution mean as CORBA, that may be delivered as a native infrastructure for the Windows NT-PC based network elements. Although it was elaborated quite quickly, it is very likely to be stabilized soon and will provide similar performances to the OMG ORBs.

TINA applications based on the TINA object model can also be implemented in Microsoft technology, although the mapping of the DCOM object model is not as straightforward as the mapping to the OMG object model. The DPE architecture itself does not prescribe the use of CORBA.

The issue of interworking the DPE with alternative distribution systems and legacy telecom architecture protocols is vital, since TINA does not intend to compete against but will co-operate with existing solutions.

3.6 COMPUTING ARCHITECTURE MARKETING

3.6.1 Benefits of TINA computing architecture

The main benefits are:

- The interoperability of the service across heterogeneous infrastructures. This point is totally achieved on the object interaction model. It is more difficult to make it fully interoperable at initialization and configuration management server level if the complete TINA architecture is not applied.
- Software durability, through portability from one infrastructure to another. The portability is more or less complete, depending on how much of the non-functional specifications of the real-time model and behavior and of the object internal specifications have been specified in an abstract way, which is not infrastructure dependent.
- The differentiation between rapid application developers which profit from distribution transparencies, and DPE runtime implementers which know how to tune a platform for real-time. When achieving this separation, and when providing tools for automatic generation of interface stubs and of objects, a large part of the development effort can be saved that can be reinvested in other service developments.

These three benefits can be partially obtained with plain ORBs, but there is a factor of gain in portability, in re-use and in application development automation for the telecommunications domain, as long as the TINA DPE is used.

3.6.2 Timeliness

- There is still no DPE profiling that would provide a minimal subset of the TINA computing model for small real-time embedded systems. This may be an issue for the small terminal or for embedded control units in small equipment. But it is assumed that the same issue will be tackled by the CORBA community in the very near future to bring industrial solutions.
- Timeliness is considered medium term, due to the use of CORBA, with its emerging toolchains which do not provide the same maturity level as the large industrial applications tool benches available.

3.6.3 Industrial strategies

As for the rest of TINA, the computing model is a source of interest for three different communities. Their strategies may diverge as their interest in TINA architecture and solutions derives from different points of view.

3.6.3.1 IT vendors

For IT vendors the DPE is an extension of their ORB products which should be an opportunity to sell development platforms and tools. A few vendors specialized in ORB software package (IONA, Chorus/SUN) are even more motivated for a rapid introduction of DPE-like architectural products. They see the DPE servers as additional offers, possibly by third-party vendors that they would mention in their catalogs in a similar way as in the complementary tool vendors' catalogs. DPE integration would thus be a different business from the ORB package sales.

3.6.3.2 Telecommunications vendors

For telecommunications vendors the DPE is a way to obtain software development automation and to find software robustness to specification evolution, due to the clearly structured ORB-based architecture. With the service architecture and the addition of a higher level of service modeling, the DPE tools create an open development framework that could serve joint development with customers: a favored requirement of operators. They see themselves as DPE server vendors, but they also expect part of the TINA middleware to be offered by third-party competitors or operators, so that integration becomes a major and challenging business. They would consider plain CORBA as a first step of product evolution towards TINA, and such products were already available at the end of 1997. In a second step, they are interested in the standardization of the middleware interfaces towards the DPE, so that they can buy the components that they have not targeted as a key technology (Figure 3.18).

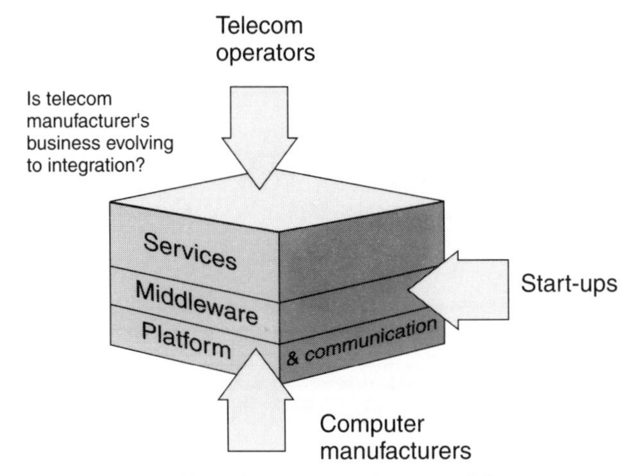

Figure 3.18 TINA DPE provides a focus on a business trend for more openness

3.6.3.3 Operators

For operators the DPE is a way to obtain open diversified platform delivery and to ensure software durability. It is also a means to specify the framework that they might use for their own development. They do not see themselves as global integrators, as the set of heterogeneous network elements and detailed third-party vendors interface configuration brings more complexity to this task. They expect strategic alliances with vendors and with other operators to solve the non-fully specified part of the computing architecture, especially interworking with legacy solutions. They would prefer to have standards or *de facto* standards for DPE servers and object model solutions in order to see the integration of components from different middleware vendors.

3.7 WRAPPING UP

The computing architecture used in telecom control and management products may rely on the standard CORBA computing products, as long as they incorporate the telecom extensions expressed in the OMG telecom group by the TINA members. In that sense you could have been spared reading this chapter.

But telecom product architects will also probably build on top of CORBA the telecom middleware servers and libraries that provide lifecycle conditions, real-time and quality of service control, product configuration and all the required foundations for an open software development environment. These include telecommunications service architecture generic blocks and resource management generic objects that will be described in the two following chapters.

In fact there is a major opportunity for a new market of such middleware products, and TINA provides guidelines on market segmentation and top priorities. For those start-up companies which engage early in the business of telecom DPE middleware and software engineering there will be excellent profit. The world market is estimated at about $200 million for five players and less than 20 customers, a typical scale for a good profit.

Chapter

4 Service architecture

4.1 BACKGROUND

In Chapter 2 the scope of the service architecture was defined in relation to the resource architecture. Figure 4.1 will remind the reader that this part of the architecture aims to define an infrastructure for the needs of consumers, retailers and third parties. "Third party" covers a wide range of potential business activities in the overall business model. This was envisaged to be on/off-line content provision and supporting delegated service logic, although other useful specializations of third parties supporting dedicated functions such as mobility support became apparent.

Until the introduction of the business model the service architecture was formed around the view of consumers and service providers. This was seen as untenably simple and the concept of service provider needed refinement. The entire Consortium did not have a unanimous view, with members weighing differently the various world trends. Ultimately, the service architecture endeavors to take the best of the Internet and telecom paradigms and leave the worst behind. For example, the service reference points do not assume any particular size of enterprise, thus emulating the low entry threshold and rich diversity characterized by the Internet. Furthermore, TINA enables more complicated relationships among service providers and consumers (noting that subscribers and end-users may be involved) and takes into account traditional telecom considerations such as user and mobility.

Certain pitfalls have also been avoided – the idea that information services need intelligent networks (a tunnel vision of many: but poor internetworks are very good for many purposes) or that services can deliver quality streams without some network management (an occasional fantasy found in the Internet world). The reality is between these

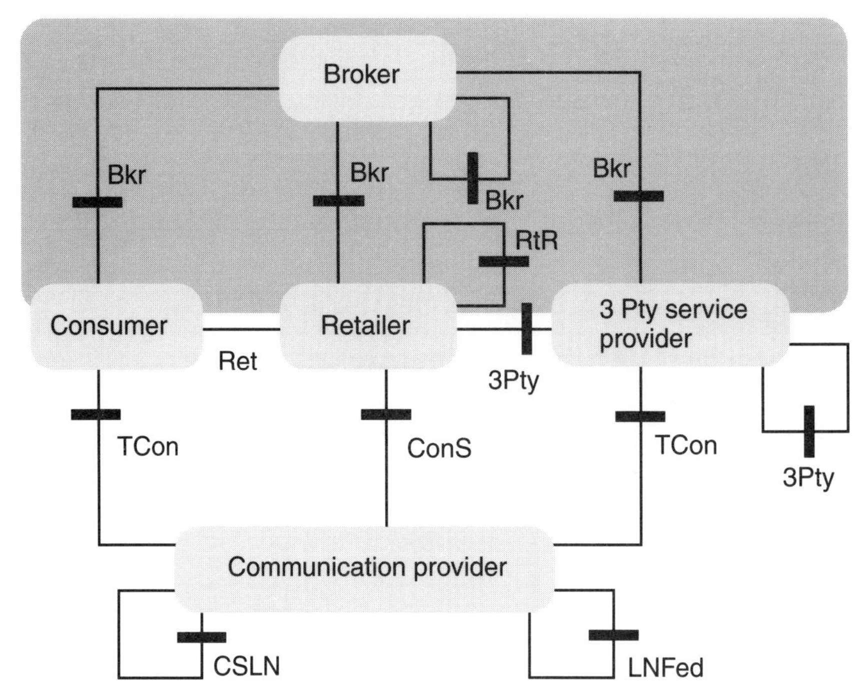

Figure 4.1 Scope of service architecture

extremes. There is too much capital, technology, and penetration of existing technology in the Internet, POTS, IN, ISDN, LAN, GSM to make these obsolete. TINA attempts to bring a common approach to obtaining access and use of services which are more than bit delivery (certainly more than bit delivery in a restrictive encoding like 64kb/s speech).

There are obviously many facets to the relationships among these business roles and the service architecture attempts only to "standardize" limited aspects relating to setting up and controlling the lifecycle of on-line instances of non-specific services. One difficulty TINA has experienced is the generality of its notion of service. In principle, this could be a service to domestic consumers (e.g. video-phone, home-shopping, conferencing) or a service between retailers for service support (e.g. the video stream servers or static content). From TINA's perspective all these examples are discrete services and any consideration of the content is beyond our scope. The service architecture addresses the infrastructure into which old and new services can be organized.

In the following sections the main features of the service architecture are described. The reader will find that the ensuing discussion emphasizes realizable implementations. Many early TINA concepts, as befitting research, were theoretical and their implementations frustratingly elusive. The aim here is to cover some of the practical and interoperability issues of primary concern to designers, drawing mainly on the later parts of the consortium's output.

4.2 SCOPE OF THE SERVICE ARCHITECTURE

Because of its background mentioned above, the service architecture focuses on schemes needed for consumers to obtain services from retailers, and retailers to obtain services from other retailers and third parties. The architecture tries to generalize specifications across these schemes emphasizing commonality. They are simply providers that need a service framework in order to offer services, whatever type they may be. Thus a mobile, domestic user of retail services is not distinguished from a retailer accessing a third-party provider. In reality, the implementing technology may be different because the transaction rates and types may be different – "may" because the enterprise's size is undefined. Generally, the needs of a one million customer per hour retailer will be different from a third party with ten customers especially if these make a million service requests per hour. Geographic distribution and session holding times will be different. The difficulties in creating an open widely applicable architecture for enterprises with such different scale profiles has meant that specifications are constructed with a minimum of specific engineering assumptions. This is a strength and a weakness. If there is a design bias, it is towards numerous and small rather than few and large. Current hardware and software trends for numerous cheap process units are more credible investment patterns for growing providers than a few monolithic tera-flop systems so favored in traditional telco OSS. Having said that the service architecture concentrates on service provision it is helpful to delve inside TINA's perspective of a service provider's internal organs.

Figure 4.2 illustrates a simplified view of a provider's infrastructure, required to host and invoke services. In the following discussion the terms "management" and "control" have a broad separation; "management" covers the maintenance of more stable contextual

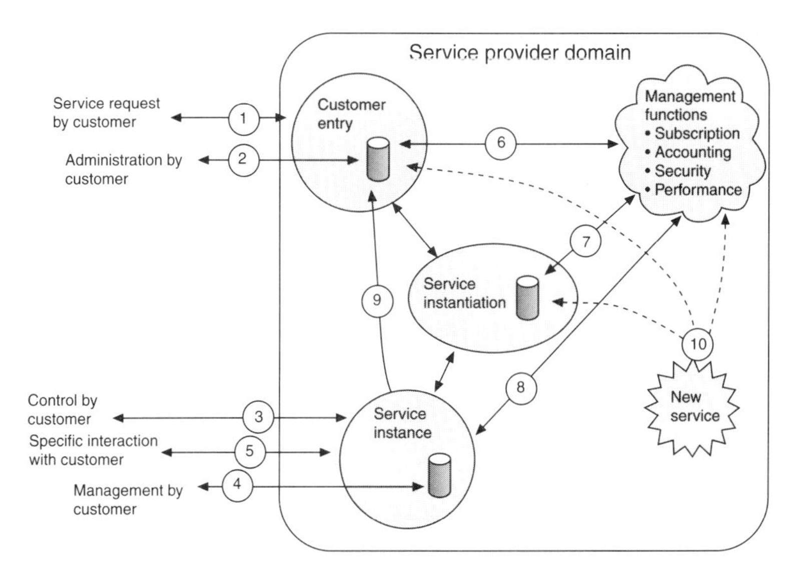

Figure 4.2 Management considerations of a service provider

information and "control" is applied to instantaneous modification of system behavior in relation to current system events.

The reason for exploring this is to consider where the architecture gives real solutions, and where there is just wishful thinking. In the following discussion the word "customer" is found; it is deliberately chosen to be ambiguous as to whether the customer represents a consumer, a retailer or a third party, because any of these may seek a service from a provider. In this sense the discussion addresses issues at the consumer–retailer, retailer–retailer, retailer–third party and broker reference points.

A provider's management and control considerations fall into four main categories and the numbers correspond to those in Figure 4.2.

- *Customer entry* These components support the customer entering or being invited to enter the service environment under managed permissions, constraints and optional preferences – these are described in the customer's profile. Such subscription profile information is at the heart of the service architecture for personalizing a customer's experience because the profile is present irrespective of the customer's terminal or location. The subscription attributes must be administered by the provider (6) (to create, modify subscription details) and the customer (2) either as the subscriber (e.g. a household parent or corporate administrator) or the end-users (e.g. children, corporate staff). One feature in TINA is that subscription administration by customers (2) may be performed through the service itself (if it relates to a specific service) or otherwise via dedicated administration services. This results in the interaction between the entry components and service components (9). Thus TINA interactions (2) are therefore also a type of the service-specific interactions labeled (4), although some simple registration procedures for mobile users are retained at the access level. Finally, to make money the provider must allow the user to view available services and start them (1) – availability may be limited by the customer's current location or terminal. Another important part of interactions (1) is that of the provider delivering service invitations to the customer – analogous to the telephone ring but much more functional.
- *Service instantiation* This area covers the management of information (6) that determines where and how a service type should be instantiated and initialized. These components are responsible for taking a service-type instance requested at the access part, locating a factory capable of launching executables and initializing the instance. This management activity is an obvious source of competitive edge for a provider domain because it determines how efficiently resources are exploited, particularly where dynamic optimization occurs. For example, it can optimize trade-offs between network congestion against CPU distribution to move a service instance process closer or further from its users. It is clearly related to the management of the providers' overall object and DPE node system. Because of the strong dependence on DPE technology, and commercial differentiation this area is addressed only in a limited way, to resolve interactions between the customer entry components and service instances.

- *Service instance management and control* Service instances are naturally the main objective of the architecture. Having successfully started the service a customer will obviously want to control (3) its lifecycle. This is called (service) session control and includes ending or suspending a session and inviting others to participate. In many services there will be non-standard service specific interactions (4) (e.g. changing the homicidal tendencies of monsters in a combat game). A third class of interactions with the consumer domain are those of contextual management (5), particularly relating to confidentiality or accounting (e.g. on-line billing). Such management contexts are closely tied to the service specifics occurring over (4). Generally, these are not interactions that human customers would directly effect, but rather they would be used by the components to manipulate and exchange common management information over traditional FCAPS functions. Finally the service provider may have a number of interactions with services instances (8) to exert control over their lifecycle (e.g. resource monitoring, fault recovery and crime prevention). Other interactions (9) may occur between service instances and customer entry (e.g. to notify of usage or accounting) in addition to the subscription profile changes discussed above.
- *Service-type management* Introducing new services is a traditional bane for telcos which hold a large installed base of inflexible, hardware-oriented technology. Often the challenge of integrating new with existing services has been solved by avoidance through adding overlay control networks and OSS. Relieving this affliction was a major motivation for TINA-C. The architecture considers the new service-type (or versions) lifecycle which moves from installation, activation, inclusion in subscriptions to eventual deactivation and withdrawal. Managing this lifecycle is the key differentiation of a service enterprise; a technically uncompetitive service platform increases cost or launch dates of new services. The best QoS and brand loyalty in the world will not compensate entirely for this handicap. If the new services are to be integrated easily then the supporting service management infrastructure must be kept stable for existing services. This implies that an enduring service support framework should support "sockets" and "plugs" internal to the provider domain, which arbitrary, conformant services could interconnect. The sockets and plugs translate to the interfaces that are expected and supported by both the service and host framework components. Additionally, an analogous supporting infrastructure could be needed in the customer domain if this also contains parts of the service support framework. Thus a major factor in the proliferation of TINA services may depend on the standardization of service-to-service infrastructure interconnection, internal to the provider and perhaps also the customer domain.

There are several issues that are not illustrated in Figure 4.2 but which are considered outside the scope of the service architecture. Most of these belong to the realm of the object management middleware and are common to an entire object system. They cover aspects such as failed object recovery and exception handling, kTN performance, node performance, load balancing, and object distribution. The service architecture excludes these issues, leaving them to clever ORB implementers to solve, allowing service creators to concentrate on service innovation.

4.2.1 Application of service architecture concepts

Overall, the TINA service architecture has good concepts and information models that apply to interactions (1)–(6) and (8)–(9) which could be used in any interface specification (TINA or proprietary). The architecture has, at the time of writing, proposed computational models including interface specifications to support (1) and (3) that give uniform access and service session control models – these are described in detail in the following sections. The computational models have been applied to give interface proposals for interactions (6), (8) and (9). Many of the computational issues affecting (6), (7) and (8) border on provider-specific issues which have achieved variable levels of detail. The informational and computational models affecting new services, (10), are, as previously mentioned, highly proprietary and have no tangible specifications.

What follows is an overview of the main concepts that have been established by the service architecture and are formative to the specifications and to the input to other fora influenced during the lifetime of the Consortium Core Team. The discussion deals with orders as basic session concepts, then the information models within the service session and finally computational issues. This is an empirically established means of presentation, not necessarily a purist view on information modeling (but on this there are usually as many irreconcilable views as there are purists!).

4.3 BASIC CONCEPTS

The following concepts were established in Chapter 2:

- The separation of access, service and communications
- The classification of services into ancillary and primary
- The fundamental sessions: access, service and communication sessions.

This section will say more about the ancillary and primary distinction, sessions in the service architecture and interaction roles. Although discussed in Chapter 2, they are revisited in detail and in the context of the previous discussion on provider domain issues.

4.3.1 Access and types of usage – ancillary and primary

A distinction is made in Chapter 2 about different types of service – ancillary and primary. It was noted that the term "ancillary" applies to a service that enables customers to modify and set up their subscription profiles (e.g. where invitations are delivered, address books, billing options and, end-user rights). This type of service has various names according to the forum, but typically the terms "management" or "administration service" are heard. TINA is unusual in that it treats management services as any other type of service usage despite the fact that they modify access-related objects. The rationale is worth explaining.

Like any other revenue-generating service, management services are an obvious commercial differentiation among providers, ranging in complexity, pricing and communication needs. The similar architectural treatment gives providers opportunity to install bespoke management services on the same infrastructure that supports all other services. An additional benefit is that the access part contains only minimal, essential features to find out about launch services and register mobile users – akin to lightweight signaling. This keeps access interactions provider-independent and consistent with the notion of universal, terminal independent access. The distinction between ancillary and primary services is retained to separate the management services from regular telecommunication information services.

4.3.2 Session models

The service architecture is primarily concerned with the access sessions and service sessions – these affect the Ret, RtR and 3Pty reference points and intra-domain interactions. The communication session is important but only from the perspective of knowing how it interacts with the service session. Note that these sessions may span more that two domains. A complicated service may include several consumers, a retailer and a third party. In these cases the session concepts are important because the distributed components share a common purpose and management context.

Figure 4.3 illustrates sessions in an example involving three business roles, a consumer, service retailer and another party which could be a consumer, third party or retailer.

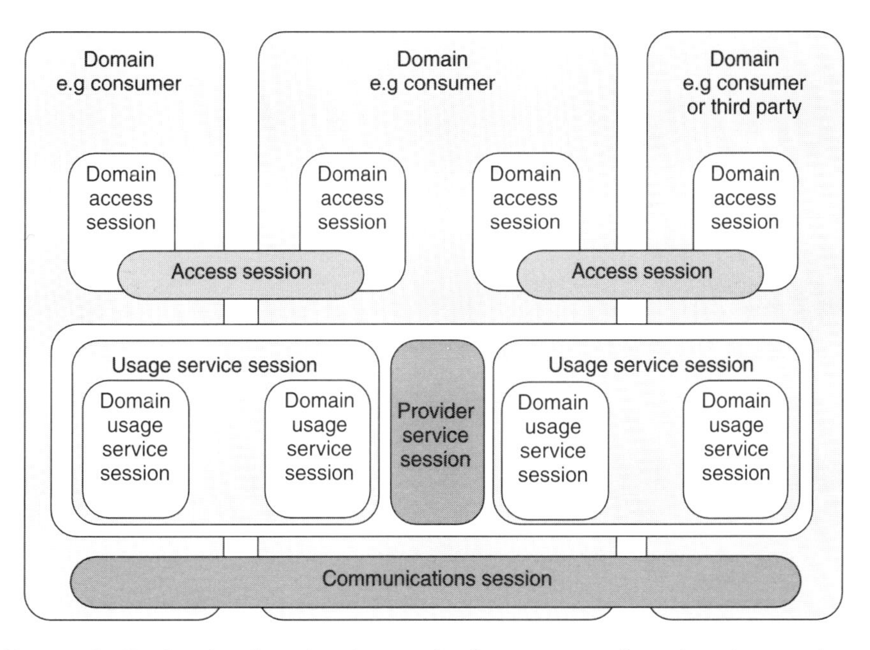

Figure 4.3 Sessions in a three-domain example of consumer, retailer and another party (consumer or third party)

There are two abstract session types: those that are confined to one domain and those crossing two or more domains. The interdomain sessions can be regarded as the relationship among the domain sessions. The sessions will now be explained.

4.3.2.1 Access session

This models the information describing the access relationship between two domains that are bound together in a secure association, described in the OMT diagram (Figure 4.4). An access session becomes established when two domain access sessions (in different domains) agree rules for the interaction. The domains may be strangers to each other or may have a long-established contract. If they are known to each other then one or both of the domains may hold established subscription information such as the constraints and preferences of the other domain. If the domains are newly introduced then such data must be assumed by default. It is clear that an access session is a prerequisite for one domain to give another entry to its resources, providing protection for both participants from each other and outsiders. The establishment of this session is inseparably tied to ORB security and key distribution technology which was covered in Chapter 3.

The lifetime of an access session is undefined in TINA. Clearly it can only exist when both its domain access sessions exist. Other than this limitation, its lifetime is dependent on the trust that each domain places on the security technology and commercial value of the services involved. For example, the access session lifetime between a provider and a user of low-value services from a secure domestic premise could be treated less restrictively than a session allowing high-value services to be invoked from a public terminal – even if the same person is involved. Similarly, the access session between a large retailer and third party may be stable over long periods of time.

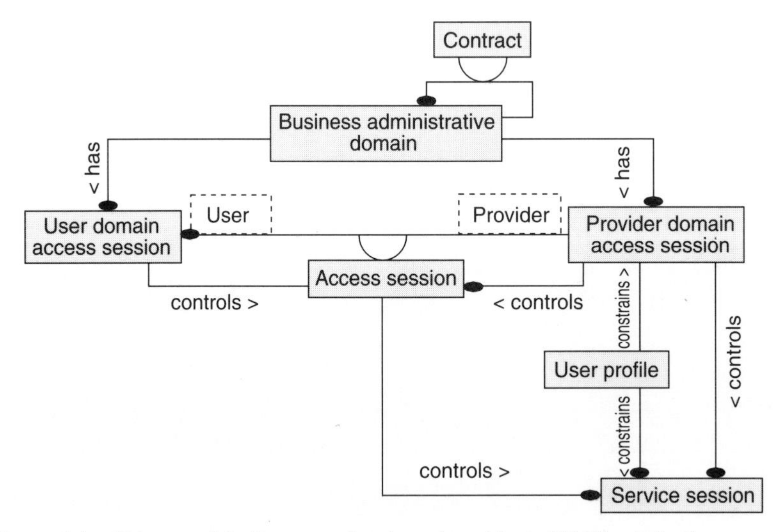

Figure 4.4 Object model of access-related session objects (OMT notation)

4.3.2.2 Domain access session

This is confined to a single domain and represents information about the home domain and other domains with which it may interact. Thus a long-term relationship between two domains implies two domain access sessions, each holding information about the conditions and capabilities it has to establish an access session. The domain access session is specialized according to predefined access roles which are discussed in section 4.3.2.1. The lifetime of a domain access session is as long as there exists capability to form an access session.

4.3.2.3 Service session

This is the most prominent session, and subject to the majority of TINA information modeling. The service session represents the information and functionality to execute a service, whether it is a single-user information retrieval service or a complicated multiparty, multimedia conference. Because it encompasses all the objects and resources necessary for a service it also spans all supporting domains. It is important to appreciate that the service

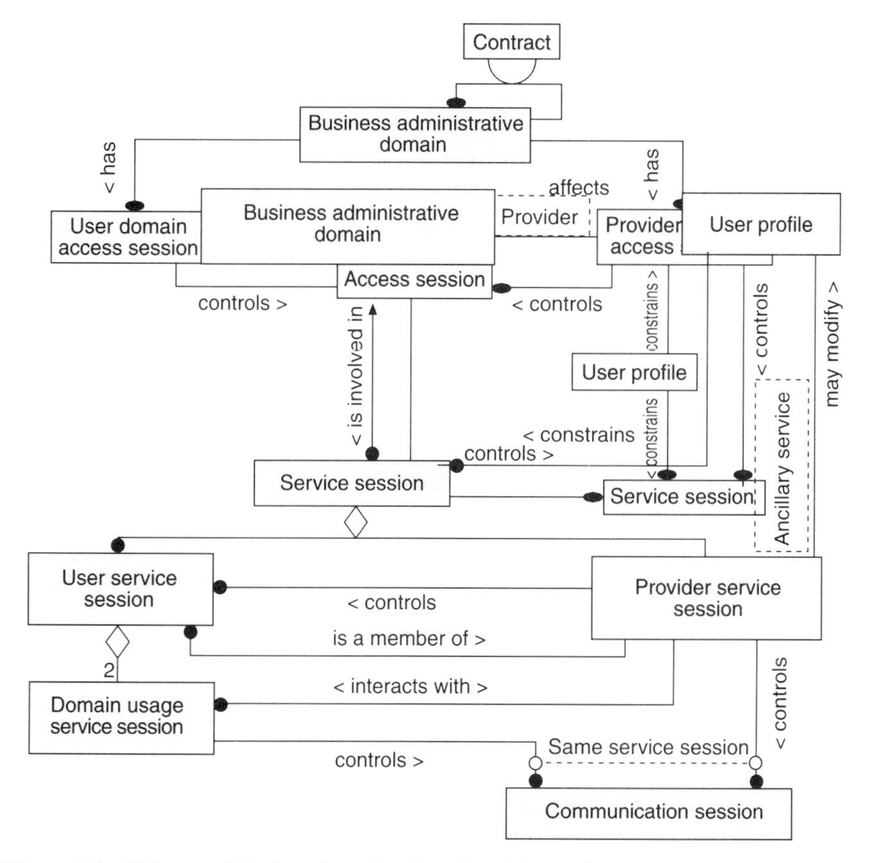

Figure 4.5 Object model of service-related session objects (OMT notation)

session represents a single instance of usage of a service-type instance; if it were compared to a conventional POTS service its lifetime would be the duration of the call.

Service session logic can be distributed among components in a service according to any feasible design rules, including dynamic redistribution during the session lifetime. Typical parallels are to Internet-based conferences where the conference logic is distributed entirely on the users' terminals and particular functions like chairmanship can rotate. However, although such anarchy is definitely not excluded, the TINA session model presumes that there may be centralized core logic to the service with which multiple members interact. The aim is to explicitly support managed services that may be performing a third-party stream set up, and embrace other related technology such as network computers. The service session is therefore considered an aggregation of two other sessions, the provider service session and usage service session as shown in Figure 4.5.

4.3.2.4 Provider service session

The provider service session represents the core logic of the overall service session, maintaining knowledge about all the members of the session and their relationships and contexts. It may also be responsible for stream connection set-up, irrespective of whether it supports any stream interfaces itself. This session may have a minimal amount of service logic but it does represent the focus of management for the overall service session, even where this is largely connection-oriented service but where events such as account management are necessary.

4.3.2.5 Usage service session

The usage service session embodies the presence of a member of the service session interacting with the service's core logic in the provider service session. It comprises two interacting domain usage service sessions, one in the provider domain and the other in the session member domain. There are two main facets to the usage service session. First, it insulates the service core from member-specific technology and, second, it allows the service core to function independently of a particular member's activity. The usage service session could incorporate different executables according to the application type or version used by the member in their domain usage service session. In principle, this gives a degree of re-use of the provider service session where its functionality is unaffected by evolution of the members' applications. It may be possible for a member to suspend participation and resume later (having perhaps moved to a different terminal) with the provider service session being unaware or minimally affected.

It is too early to know whether such separation at an object level of the provider domain usage service session and provider service session components has real implementation and re-use advantage. For many services, both sessions will be implemented by the same component, conceivably within the same CPU process. In these cases the advantages are only realized by service designers through elegant object design and programmatic APIs that can be applied to a family of services. Where separate provider and usage service session components are distributed between processes or servers there may be additional benefits to the service operator depending on service complexity.

4.3.2.6 Domain usage service session

This session covers the domain-specific information relevant to the membership of a service session. As such the domain usage service session must hold at least default information about the capabilities of its own domain and the restrictions, constraints on the other domain usage service sessions with which it is bound to form the usage service session. The domain usage service session is specialized according to predefined service roles. This is analogous to the domain access session which is specialized according to access roles.

4.3.3 Roles

The final, basic concept is that of roles applied to the participant in an access session and a usage service session – sessions which are explicitly interdomain. These sessions embody a more protracted and complicated relationship than typical client–server synchronous transactions and therefore the interfaces that domains export to each other are determined by the business of each domain and the role it wishes to present within each session instance. The service architecture proposes two pairs of complementary session roles: user–provider and peer–peer which both apply to the access session and usage service session. There are, however, slightly different semantics for each role in the different session types.

User–provider is designed asymmetrically and presumes in the access session that the principal in one domain is requesting service from the other – typical of many interactions such as browsers on the Internet. In the usage service session the provider role is associated with the domain supporting the core service logic for that service session. The service session user role is associated with membership of the service. While it seems intuitive that the access-user role results in the same principal playing the service user role, there are good reasons for not presuming this. It must be understood that an access-user is alternatively regarded as simply requesting provision of object services. There is no architectural constraint that those requested object services must conform to TINA or, if they do, that any particular service role must be supported. It is important to see access roles and service roles as independent, even if they are identically and confusingly named.

However, there is a caveat which will be obvious for implementers of some advanced service features such as "suspend". For example, a domain offering a "suspend service session" operation as part of its provider role in a service session will also need to offer the complementary operation "resume service session" as part of its provider role in the access session. Otherwise the hapless user could suspend a session without any obvious means to resume it. Thus coupling resulting in some provider-role features at a service level (most notably the suspend service and suspend participation) must be associated with provider-role features at an access.

The peer–peer roles presume that each domain's exported interfaces are a mirror of the other's. In some respects this role at both access and service levels is a combination of user/provider roles. However, a distinction exists because the peer role represents shared information behind the composite user/provider interfaces in one domain. Hence, one peer–peer access session can support service requests in either direction. In a service session, complementary sides in a usage service session support provider role interfaces and, for example, a suspend service sent in either direction affects the overall service session albeit in different ways. More detailed relationships are described in the TINA Service Component Specifications Part 1.

4.4 SESSION COMPOSITION

This part of the architecture explains what sessions are present and how they are related when multiple domains have components contributing to a service session. This is potentially a key area to enable management and control of services across multiple provider domains at the service (not connection) level. The concerns are primarily those of simultaneous involvement of consumers, retailers and third parties. Because the primary TINA building block of management and control is the session, these composition designs are constructed using sessions rather than components. The architecture attempts to re-use a small number of session types in a variety of different configurations, and because of the containment relationships (e.g. a D_USS inside a usage service session inside a service session) some multidomain configurations rapidly become diagrammatically complicated. This section discusses only two probable service configurations and more esoteric examples can be found in the architecture documents.

The service architecture makes a distinction between three concepts; domain federation, service federation and service composition. Domain federation is a prerequisite for service federation and service composition and amounts to a statement that domains must agree terms to support service aspects across domains. Federation is nothing new to network operators which have, for many years, employed inter-operator signaling and connection gateways for services such as international direct dialing and GSM. The means to establish and maintain federations is assumed in the TINA architecture. The distinction between service federation and service composition is simple. Service federation reflects mutual arrangements to operate services and therefore covers aspects such as the operation of compatible services, invitation delivery across domains and service sessions support across multiple domains. The last is referred to as service composition and is therefore a subset of service federation. Some aspects of service federation such as delivering invitations to users in other "home" domains affects the reference points and is specified there. Note the examples below consider only service composition.

4.4.1 Example 1

This is the first, most intuitive configuration and can be seen in Figure 4.6. A service requires the involvement of a third-party service provider as a participant rather than providing core logic. This case is analogous to premium rate POTS or IN services (e.g. the so-called "chat-lines" irresistible to some teenagers – but the bane of their parents). Here the third-party provider is represented as another participant similar to the consumer but only in respect of the degree of control over the service core, although the commercial relationship is clearly different. In the general case the third party's financial contribution is not the same as the consumer's but the underlying application components may not differ. The session control is comparable for the third party and consumer, although the exact features available or enabled may differ (e.g. a user may have the capability to invite another user but not the third party). The important message here is that while each usage service session has common control features (e.g. endSession operation), the business relationship between the consumer and third party may be handled outside or inside the session according to the service designer's choice and inclusion of service-specific interfaces. This makes garage-sized service content provision an open market with a flexible division of responsibilities (e.g. if agreed, the retailer manages the service, communication session, consumer billing and accounting). A final important note here is on separation of access and usage roles discussed in section 4.3.3. The example has tacitly assumed that the third party becomes involved in the service in the same way as another consumer, through the delivery of an invitation outside or inside an access session. This does not have to be the case. The separation of the roles in the access part and usage makes it possible, provided the retailer and third party both agree to this, for the retailer to request that the third party initiates the usage components necessary to participate in the service. This does not affect the role that the third party takes in participating in the service session.

4.4.2 Example 2

Figure 4.6 illustrates how a retailer might sub-contract the service core logic to a third-party provider, but with the retailer retaining parts of the management or session control

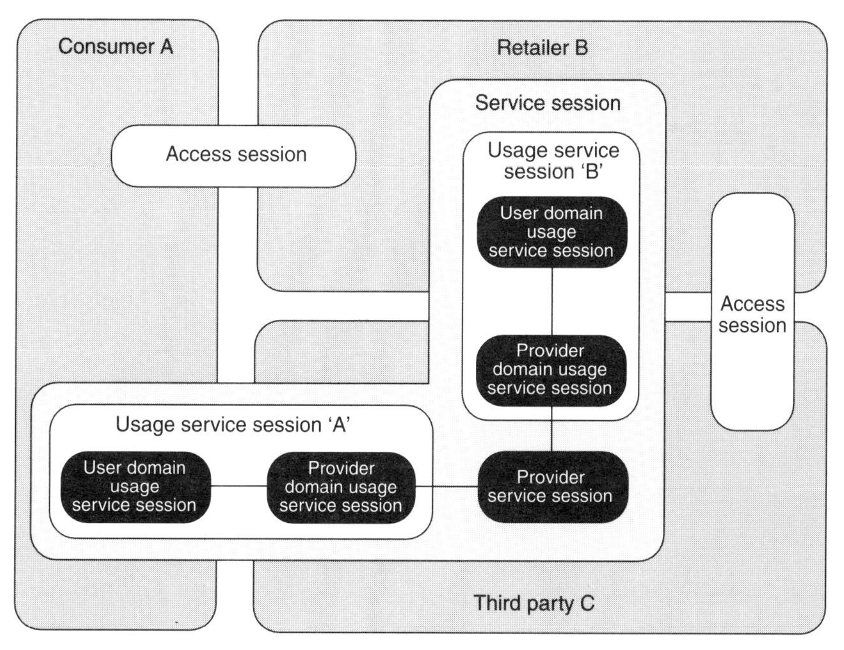

Figure 4.6 Example of service composition showing service logic sub-contracted to third party

of service session. For example, the consumer requests a video on demand but the source of the video is a third-party provider who also offers the user the video selection and stream control interfaces (to pause, play, stop the video). The consumer may not see the third party as a separate domain since at an application object level only interface references are a concern and domain location of these is largely unnoticed; DPE security does need to be aware because secure associations must be delegated across domains. The consumer sees one perspective of the session through usage service session "A" – that appropriate to an end-user consuming service content. However, "B", the retailer's view, is that of session management and high-level session control; assuming the retailer has no interest in interaction with service-specific content. These interactions are part of the 3Pty reference point but basic session control is common with Ret.

There are obviously a variety of other configurations that can be imagined. Many are explained in the service architecture document and involve the specialized domain usage sessions such as peer and composing sessions. These allow two service sessions to interact in a variety of ways, for example enabling the core provider of one service session to initiate a subsidiary service session in another domain to obtain service content. In these situations it is possible to arrange many consumer–retailer service sessions all using a single retailer–third party service session to obtain specific content. The choice of arrangement is decided by the retailer and third party, but is ideally invisible to the consumer.

4.5 INFORMATION AND COMPUTATIONAL SPECIFICATIONS

It was previously stated in the definition of "session" that component objects of a session share common information and context. The TINA service components specification and reference point document aim to specify first, what information is held by the various sessions (the information specification) and second, how that information is manipulated and exchanged internally and externally by the component objects of the sessions (the computational specification).

In the following sections some specific information models covering the service session will be discussed. In addition to the service graph there are management contexts and management context interfaces which address some of the FCAPS needs of service session elements

4.5.1 Information models

4.5.1.1 Session graph

The aim of the session graph is simple, to maintain a knowledge of the state of a service by holding the entities that are members of the session, their permissions to change the session state and the stream bindings in which members are involved. The heart of such a model is shown simplified in Figure 4.7.

The graph is an aggregation of two abstract types, the session member and session relationship, which can be associated in many ways to build the picture of entities involved in a session and what they can and cannot control. Session members are specialized into

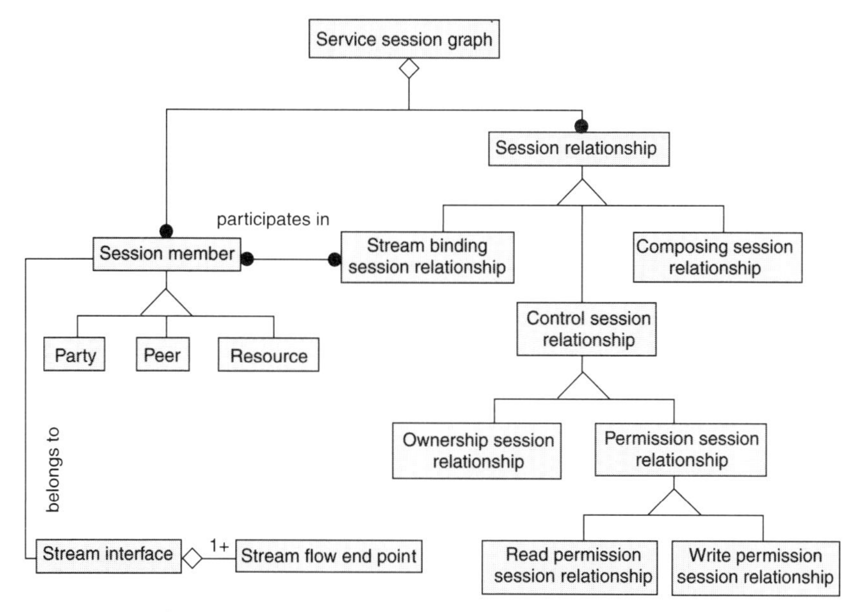

Figure 4.7 Service session graph object model (OMT notation)

classes whose instantiations are represented in the usage service session part of the service session, party, resource and peer. The relationships these members can hold are limited to basic types suitable for all services, stream binding, control and composing. Although more detailed definition and some other features of the service session graph can be found in the deliverables, it is worth explaining some basic aspects. The session members describe the participation in the service and there are three types:

- Party represents a service user in a session and tags information about that user's participation. A user effectively becomes a participant in the service only when instantiated as a party member. Parties may also invite other parties (users) to a service session. The permissions and constraints that a party member has are defined by the control relationships. The party is the only member that is found in every service feature set. Parties can by nature be active in services by modifying other parties and resources.
- Peer is important to composition relationships where one session participates in another session, and they are both capable of exerting session graph control over each other. The inclusion of peers is a service logic decision because it represents a more complicated view of service function that is shielded from parties. A peer would normally experience only the composing session control relationships because it represents an association of sessions or a mediation object between two equal partner sessions.
- Resource models supporting elements for the service session such as a file,

whiteboard or conference bridge. A resource can be instantiated by the service logic or by direct request or involvement of parties and peers. Resources are passive and modify or prompt modification of changes to the session graph.

The session relationships should be obvious for the control type which can be parallel with UNIX-type file access permissions. The composing relationship is applied to the area of composition of service sessions, where, for example, one session is interacting in a peer role with another service session – i.e. both sessions are using the other in some equivalent respect to meet a common goal. Some of the subtleties in composition are addressed in section 4.4.

Because the session graph is oriented towards supporting services that require managed streams it has a number of features to hold stream binding information at a level depending on the abstraction the service designers find simplest to use. The abstractions are participant-oriented stream binding and flow-oriented stream binding. The difference and additional features of the session graph can be appreciated from the object model in Figure 4.8 which overlaps with Figure 4.9.

Participant-oriented stream binding is the greater abstraction and can be seen as the association of members with a stream binding relationship of a specified type and quality. The stream binding is also linked to stream interfaces belonging to each session member involved in the stream relationship. Participants supply the stream interfaces instances. The binding can be mapped onto many stream flow connections. The bindings are multiparty and multipoint to multipoint. This abstraction means that a simpler view of stream bindings can be manipulated, for example adding or removing participants and changing the type of stream by specifying a different video standard. If the service requires mapping, the next level of detail can be used.

Flow-oriented stream binding maps the stream relationship to the stream flow connections which attach two or more stream flow end-points. A stream interface may contain many stream flow end-points. The stream binding can be considered to be the aggregation of flow connections, and session members are involved only indirectly having supplied stream interface and flow end-point instances. The fundamental difference from participant stream binding is that here stream bindings are modified by acting upon the

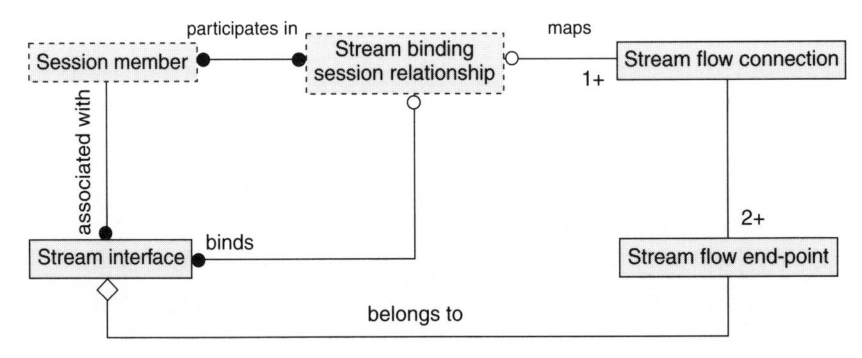

Figure 4.8 Stream binding parts of the service session graph (OMT notaion)

flow connections (a structure holding stream branch, quality and type) rather than the session binding relationship.

The feature sets that implement to the service session graph model are described in section 4.6. They build incrementally giving the service designer a choice to the extent of the graph implemented in the service core and visible to clients of the service core, in other words the participants of the service via the usage service sessions.

4.6 COMPUTATIONAL SPECIFICATIONS

So far the concepts and information views of the service architecture have been discussed. This section takes a closer look at computation and engineering issues faced in the architecture. All the information objects and roles so far covered translate into macroscopic components that may comprise co-located or distributed objects (but still within a single administrative domain). These components represent a level of TINA interoperability and therefore in their specification, as with TINA ODL, it is important to consider the interfaces supported by the component and those interfaces required.

In addition to the optimal choice of component there is the accumulation of operations onto interfaces. There is a variety of schemes to make interface design choices. Two obvious approaches are to collect operations into interfaces according to related functionality or by which client types use them. The former is useful where operations are closely related and the diversity of clients too high to define bespoke interfaces; the penalty is a potentially greater overhead of per operation access control. Tailoring interfaces to specific clients is useful where there are few client types and access control is generally at an interface level – thereby making interface hiding a useful security asset (but not a sufficient one) and eliminating the DPE security overhead of per operation access control.

The service architecture uses both schemes to accumulate operations to interfaces. All interfaces are client-type oriented but to avoid huge interfaces, functionally related groups of operations are also separated into a series of different interfaces. Each series of interfaces reveals a progressively greater part of the session information model to manipulation or examination. Therefore, a typical service architecture component must support and require some basic interfaces which confer its fundamental character, and at the discretion of the designer a number of optional interfaces up to the level of complexity required by the component.

Because many of the interactions in the service architecture are more than simple client–server exchanges, generally, each increment of capability requires interfaces on both objects in dialogue (for example, to call back, notify change, etc.) especially where those objects are supporting the complementary session roles described in section 4.3.3 across reference points. The combination of interfaces either side of a reference point which relate to a single increment of functionality is referred to in the service architecture as a feature set. Therefore, each feature set can be regarded as the capability for two objects to communicate a restricted subset of the overall, session information model they share.

It may help to reiterate that reference points represent a cross-paragraph through an object architecture at which conformance specifications can be written. Therefore, either "side" of the cross-paragraph has no visibility of objects on the other "side", only interfaces. The objects can only be deduced indirectly from a reference point specification; the inference being that groups of interfaces which have closely interdependent behavior may be supported by a single object that holds their shared information model. The point is that the TINA reference points give a very limited picture of specific interactions in the architecture and that the service components specification give the detail that fits the reference points together.

To illustrate the feature set ideas further it is worth examining the structure of the Ret reference point in Table 4.1 which shows the feature sets, their number of comprising interfaces and a description of their function. Ret is also further covered in Chapter 6.

Consider Multiparty FS in Table 4.2, which would be supported by the components user application and usage session manager. The multiparty feature set is only used where the service supports multiple users, such as a conference or video game. The complementary nature of the consumer and retailer side interfaces in the feature set is easily appreciated. Additionally, the optional interface, i_multipartyFSInfo, adds further flexibility for

Table 4.1 Structure of consumer–retailer reference point specifications

Feature set	No. interfaces user	provider	Description of feature set
	Access part (including access session)		
OutsideAccessSession	0	2	Allows a consumer to initiate contact with a retailer
InsideAccessSession	5	3	Supports fully functional access session
RegisteredOutside AccessSesssion	4*	0	Optional support to contact mobile consumers
	Usage part (service session)		
BasicFS	0	1	Very basic service session control for user
BasicExtFS	1	0	Allows retailer to request information from consumer domain
MultipartyFS	2	1	Control and information exchange for multiparty sessions
MultipartyFSInd	1	0	Additional multiparty sessions indications sent from retailer
VotingFS	1	1	Systematic rules for agreeing session changes among participants
ControlRelationshipFS	2	1	Detailed control over member permissions associated session attributes
ParticipantSBFS	2	1	Stream binding control at session member level
ParticipantSBFSInd	1	0	Supports above with extra indications
StreamInterfaceFS	2	2	For exchange of stream flow information
SFlowSBFS	1	2	Stream binding control at session flow connection level
SFlowSBIndFS	2	0	For indications of stream flow actions
SimpleSBFS	1	2	Simple specialization of SFlowSBFS
SimpleSBIndFS	2	0	Additional indications relating to SimpleSBFS

*three of these four interfaces also belong to the InsideAccessSession feature set.

Table 4.2 Interfaces of the Ret multiparty feature set

Interfaces		Description
Consumer role	Retailer role	
i_multipartyFSExe		Supports a consumer component responding to changes to its participation in a multiparty conference
i_multipartyFSInfo		Optional interface, allows consumer component to receive information to the consumer component about the multiparty status of the session.
	i_multipartyFSReq	Offers consumer functionality to invoke changes to the multiparty configuration of the service, e.g. invite another user, suspend participation

service designers to choose the support offered by the consumer application to know about the activity of other participants. For example, in a loosely coupled conference such as a lecture, participants need not know about what others are joining or leaving.

4.6.1 Components of the service architecture

As explained in Chapter 2, a TINA service component (SC) is a functional entity that interacts with the exterior via computational interfaces, offered by the SC to the exterior or accessed by the SC externally, and can be mapped onto several functionally equivalent constructs in the TINA object model. In Chapter 2 we also understood what components are and how they can be combined. Let us introduce the term service component (SC) to indicate components in the service architecture.

Service components have a certain degree of re-usability, depending on how "generic" they are. In practice, it is envisaged that some components will be absolutely generic, some will fit a wide range of services, some will be specific for given providers (constituting part of the added value of a given stakeholder's offer) or for given terminal equipment, etc. The specialization mechanism makes it possible to design "general-purpose" components and subsequently specialize them for a narrower set of applications, keeping the proper level of interoperability.

Besides defining what service components are, TINA defines a generic framework on which services run. This framework, based on the concepts explained in section 4.3, corresponds to a set of generic components, directly defined by TINA-C. Interfaces defined for these components constitute the inter- and intra-domain reference points, and ensure interoperability between different stakeholders and equipment from different manufacturers.

In some cases, TINA-C defines abstract components, that is, components that are not deployable as such but serve only as a basis to derive other component types by specialization. In fact, some service components require both specific functionality – for example, because they include features depending on specific services – and a set of generic features/interfaces. In this case, TINA defines an abstract component that comprises only the generic part. Designers will add the specific parts and thus define a subtype of the abstract component. Note, however, that specialization is always possible to give added value to any component.

4.6.2 Overview of TINA-defined service components

Table 4.3 lists the TINA service components defined by TINA-C. Components are classified into access related, usage related and access and usage related. The first column contains the full name of the component type, the second its abbreviation. The third column refers to the concepts in section 4.3, indicating which concept is supported by each component. The fourth column shows the role with which the component is associated, which determines the type of domain where it is deployed. Finally, the fifth column indicates whether a component type, as defined by TINA, is usable without any further specialization (instantiable) or requires specialization to be used. A further distinction is made for this second category: component types that are used as abstract types to derive instantiable types defined by TINA-C are described as "abstract"; component types that have to be specialized by service designers (e.g. because they contain the service-specific logic) are "incomplete".

Table 4.3 Service components in TINA service architecture

Service component name	Abbreviation	Concept supported	Stakeholder role	Status
Access-related service components				
Provider agent	PA	Domain Access Session	User	Instantiable
User agent	UA	Domain Access Session	Provider	Abstract
Named user agent	nUA	Domain Access Session	Provider	Instantiable
Anonymous user agent	aUA	Domain Access Session	Provider	Instantiable
Peer agent	PeerA	Domain Access Session	Peer	Instantiable
Initial agent	IA	Domain Access Session	Provider or peer	Instantiable
Usage-related service components				
Member usage session manager	MUSM	Domain usage service session	Session member	Abstract
Composer usage session manager	CUSM	Domain usage service session	Party	Incomplete
User service session manager	USM	Domain usage service session	Provider	Incomplete
Peer user service session manager	PeerUSM	Domain usage service session	Peer	Incomplete
Service session manager	SSM	Domain provider service session	Provider	Incomplete
Service factory	SF	Domain provider service session	Provider	Incomplete
Service support component	SSC	Domain usage service session	Provider or resource	Incomplete
Access- and usage-related service components				
User application	UAP	Domain access session, Domain usage service session	User (access session) Party (service session)	Incomplete

Components are associated with a specific role that a stakeholder assumes in a given session (both access or service session); roles are explained in section 4.3. TINA stakeholders interact using either the user/provider or the peer-to-peer paradigm, as explained is section 4.3. In inter-domain interactions, each stakeholder is associated with a role; the Ret, R+R 3Pty and Bkr reference points, as well as the TINA components behind them, are defined with this role model in mind. Table 4.1 indicates, for each component, the role taken by the stakeholder in whose domain the component resides, for the interaction the component refers to.

Below, we briefly describe the components and their interactions in two scenarios that involve all interaction paradigms: the retailer federation scenario (consumer–retailer–retailer–consumer), which involves the user/provider and peer-to-peer paradigm, and the composed service scenario (consumer–retailer–third party provider), which involves two different kinds of user/provider paradigm.

4.6.3 Retailer federation scenario

To fix our ideas, let us consider a specific scenario and use it as a basis to derive considerations that are valid in general. Figure 4.9 shows the situation of two retailers federating to provide a service to two consumers. Each consumer takes the user role with respect to the corresponding retailer, which takes the provider role; both retailers take the peer role in the relationship between them. Below, the components involved in these types

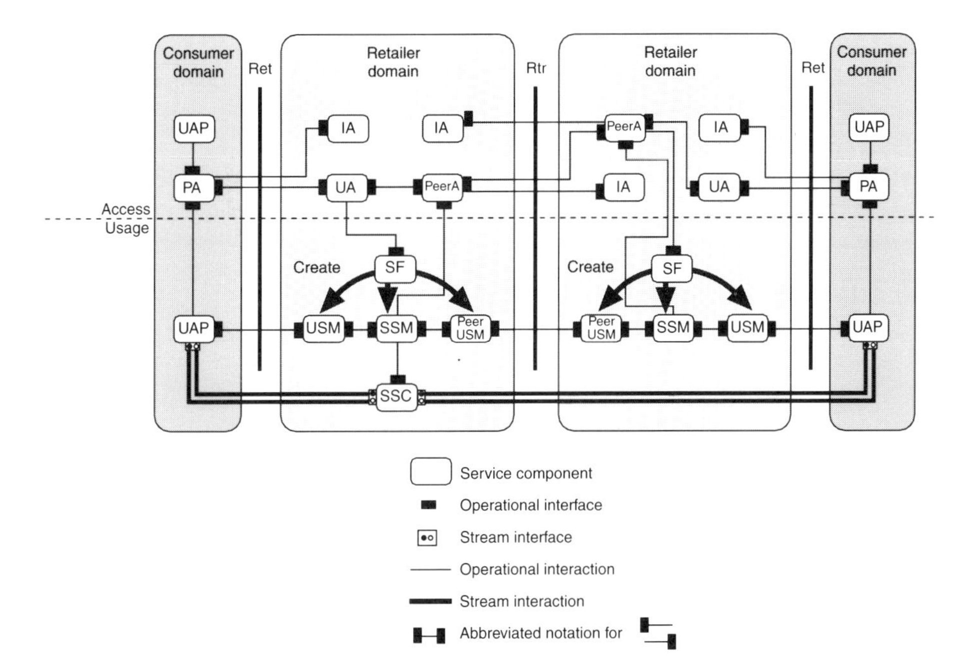

Figure 4.9 Retailer federation scenario

of interactions are described. User/provider interactions (between consumers and retailers in the scenario) are considered.

4.6.3.1 Access- and usage-related components: user application (UAP)

The user application (UAP) is a component type that corresponds to non-TINA or service-specific client applications in the terminal. As such, it is described as "incomplete" in Table 4.3. As an access-related component, no interface and no behavior are defined, these being fully specific; however, it is useful to keep the notion of this abstract component because it is the basis for encapsulating existing access applications, such as World Wide Web browsers. As a service-related component, TINA-C defines operational interfaces to control a service session from the user side, as well as the corresponding behavior. In more sophisticated terminals (PCs, network computers, etc.) there will be typically multiple UAPs, specialized for different services, access mechanisms, and so on. In simpler terminals (e.g. a mobile handset) there could be a more limited number, or even only one UAP taking care of both access- and usage-related functionality.

4.6.3.2 Access-related components

Access-related components provide a framework for secure and personalized access to services and for mobility support. In the retailer domain, the initial agent (IA) is the initial contact point for the provider agent (PA) in the consumer domain, or the peer agent (PeerA) in the federated retailer domain. The IA is used to establish an access session with the user agent (UA). The UA is an abstract component type, that is specialized into anonymous user agent (aUA) and named user agent (nUA), taking care of anonymous and named access, respectively. PA, IA, aUA and nUA are instantiable components, although they can be further specialized to include specific functionality, such as sophisticated authentication mechanisms.

The PA and aUA/nUA components interact within a secure, trusted relationship between the user and the provider (access session). They support authorization, authentication and customization of the consumer's service access, and provide a secure mechanism for starting and joining service sessions. In terms of the access session, consumers take access user roles; retailers access provider roles. It must be noted that the situation of a consumer establishing an access session with a business role other than the retailer is not prevented, although this is not described in this book.

4.6.3.3 Usage-related components

Usage-related components provide a framework for defining services which can be accessed and managed across multiple domains. Each retailer has a service session, where participants take one of the following roles: party, resource, provider and peer. Parties are the consumers participating in the session; resources are the elements that are controlled by the session, the provider is the retailer and peers are other retailers whose service session is federated with the session considered. This, of course, is the general case: in many cases some of these roles are missing, and consequently so are the associated service components. Note also that the usage role to which a component refers does not always coincide with the domain where the component is deployed. For example, Figure 4.9 shows that USMs are in the retailer domain (provider role), while they are associated with each consumer (party role).

The service session manager (SSM), user service session manager (USM) and peer user service session manager (PeerUSM) are instantiated by the service factory (SF) in the retailer domain, based on requests from UAs; service support components (SSCs) represent special resources controlled by the service session; they are static, but may become available only after the SSM instantiation. All these components together constitute the service logic. In any case, the only component whose presence is mandatory for each retailer is the SSM; also stream interfaces are always optional. Note that USM and PeerUSM are a specialization of the member usage session manager (MUSM) component, an abstract type that defines generic functionality and interfaces that are specialized according to the role. SSM and USM provide session control capabilities – the SSM supports those shared among the parties in the session; the USM supports those dedicated to each party; and PeerUSMs represent consumers associated with federated retailers in the local service session, keeping the associated "remote" session up to date with changes in the local session. The usage-related UAP in the consumer domain acts as an end-point for session control, as described above. All usage-related components are abstract types, as they need to be specialized for specific services.

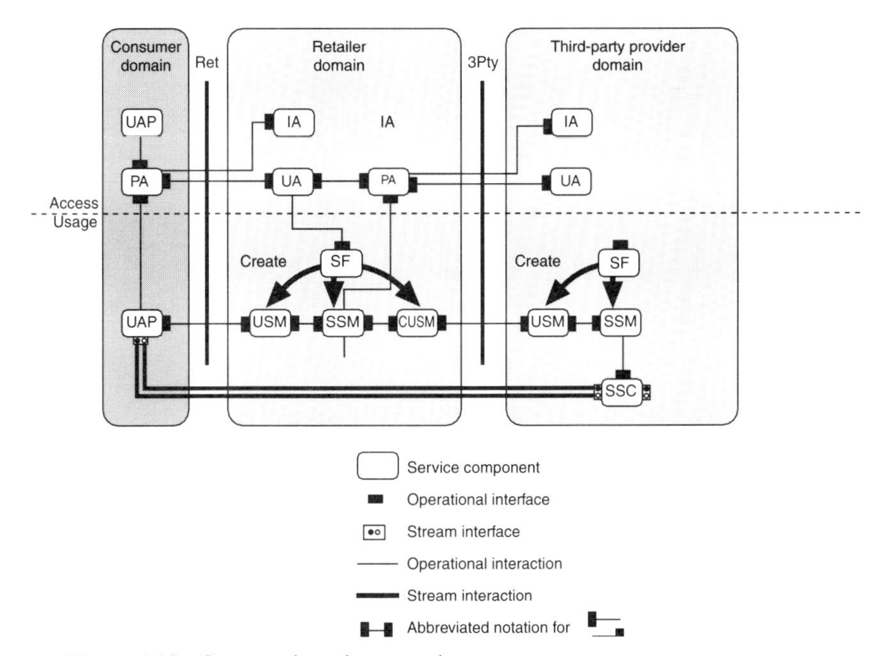

Figure 4.10 Composed service scenario

4.6.4 Composed service scenario

Let us consider a specific scenario to fix our ideas. Figure 4.10 shows a situation with a consumer, a retailer interacting with the consumer, and a third-party provider offering a service to the retailer. The service offered by the retailer to the consumer is a composition of the third-party service and a service provided by the retailer. In this scenario, both the consumer/retailer and the retailer/third-party provider relationships follow the user/provider paradigm.

User application and access-related components are unchanged with respect to the scenario in the previous section. Note that here no peer-to-peer relationship is present, while the previous scenario described both a user/provider access and a peer-to-peer access.

Usage-related components are also similar to the previous case, with the difference that the PeerUSM is replaced by the composer usage session manager (CUSM). In fact, as before, both the retailer and the third-party provider control a service session. However, while before the interaction paradigm between the two retailers was peer-to-peer, now the paradigm between retailer and third-party provider is user/provider. The CUSM is the specialization of the MUSM for a user/provider usage relationship, targeted to service composition. The third-party provider sees the retailer as a party in its session, and assigns it a USM in its domain; the interaction with the third-party provider session from the retailer side is handled by the CUSM. The CUSM is an abstract component type, which needs to be specialized with the composition logic.

4.7 EXAMPLE

This section illustrates the key service architecture concepts by telling a short "TINA story". The aim is to explain the service architecture information model, the behavior of TINA service components and their interactions examining a specific case, based on a real deployment scenario.

The scenario shows how different services, which can be provided individually on non-TINA systems, different transport technologies, existing independently of TINA, and relationships among multiple business roles in the telecommunications arena, can be managed by a single framework: the TINA service architecture.

4.7.1 Scenario description

4.7.1.1 The "story"

Franz wants to go on vacation with some friends, and decides to look for offers on the Internet. He first contacts his retailer, Supermedia Corporation. In fact, his retailer offers a wide range of services, both TINA and non-TINA, including Internet access and browsing (in the same way as a normal Internet service provider). Supermedia offers unified access to all his services, unified customer management (subscription, profiles, accounting/billing) and unified access to multiple transport networks by adopting the TINA solution.

Once logged on to his retailer, Franz browses the Internet and reaches the site of a travel agent, Gulliver Travels; he navigates through fancy vacation offers and, from time to time, he downloads graphics and views short video clips. At a certain point, he finds something interesting: a two-week trip to Cuba for $800 per person, everything included. He decides to talk to a sales representative to get more information on this offer. He is also subscribed to a multimedia conferencing service with Supermedia and Gulliver Travels is subscribed to the same service with the same retailer. Therefore the retailer enables a button on the HTML page that activates a video-telephony call to a salesperson at the travel agent. An audio and video connection is established. Franz discusses the offer with Sarah, a saleswoman describing all wonders of the tropical paradise. From time to time, video-clips add on to the voice and image of the salesperson. Franz can control the information stream: stop, accelerate, view again. The offers seem really good, but Franz cannot decide alone. He invites his friends Hiroshi, Barbara and Tanja – with whom he wants to spend his vacation – to the multimedia conference. He briefly explains his discovery, and has the saleswoman repeat the key information; they also decide to view again the main video clip.

After a brief discussion (while the saleswoman is momentarily excluded from the conference), they decide to book that vacation. Since each of the four pays for themselves, they all access the electronic payment service provided by the retailer and, after the necessary supplementary authentication, submit the payment.

Finally, they go on their trip to Cuba; this last part is not supported by the TINA service architecture.

4.7.1.2 Assumptions

Franz, a private end-user, already has a subscription with the retailer for a set of services; this includes Internet access and browsing, information stream reception, multimedia conferencing and electronic payment. Franz's friends have a similar kind of subscription. The travel agent is subscribed to the same retailer; if it had been another (federated) retailer, component interactions would have been more complicated, without any substantial difference. The type of subscription the travel agent has is different (for instance, it owns a Web server and outsources to the retailer the collection of bills).

The terminal equipment Franz and his friends are using is a personal computer equipped with multimedia capabilities (microphone, speakers and a video camera). To fix our ideas, we can assume that Franz's terminal communicates with the network via IDSL (ISDN digital subscriber loop): this gives the possibility of having multiple transport channels, to be used both for stream connectivity (i.e. interactions via stream interfaces) and for DPE-based control (i.e. interactions via operational interfaces); the IDSL technology enables sufficient bandwidth for a multimedia information flow exchange on top of an ISDN access channel. Note that, besides QoS considerations, the scenario is independent of the transport technology (both in the access network and in the core network). We may also assume, for instance, that Franz's friends are connected via plain ISDN, or even with a direct ATM connection; the only difference would be in the native communication software and hardware in terminals, with no impact at the service level.

Furthermore, it is assumed that the software for a TINA DPE client, a TINA provider agent (PA) and an initial TINA access session-related user application (UAP) are permanently available in the terminal (for example, on a local hard disk). In particular, the initial UAP offers a basic user interface and is capable of interacting with the PA to access the retailer. We can also assume that the travel agent has a multimedia-equipped workstation directly connected via ATM to the network.

4.7.2 Scenario phases

4.7.2.1 Access to the retailer

4.7.2.1.1 Local logon and initial contact with the retailer

Figure 4.11 shows the scenario with the stakeholders involved, the components and the different steps of component interactions. In this phase, two stakeholders are involved: Franz, taking the consumer business role, and Supermedia, taking the retailer business role. First, Franz sits at his PC and, if necessary, logs on to it locally. Then he accesses the Supermedia retailer in order to access the service portfolio he is subscribed to and starts with World Wide Web browsing. Franz inputs the name of the retailer he wants to access via the user interface of the initial access-related UAP (1). The UAP forwards the information to the PA (2), which contacts the initial interface for the retailer (3): this is an oper-

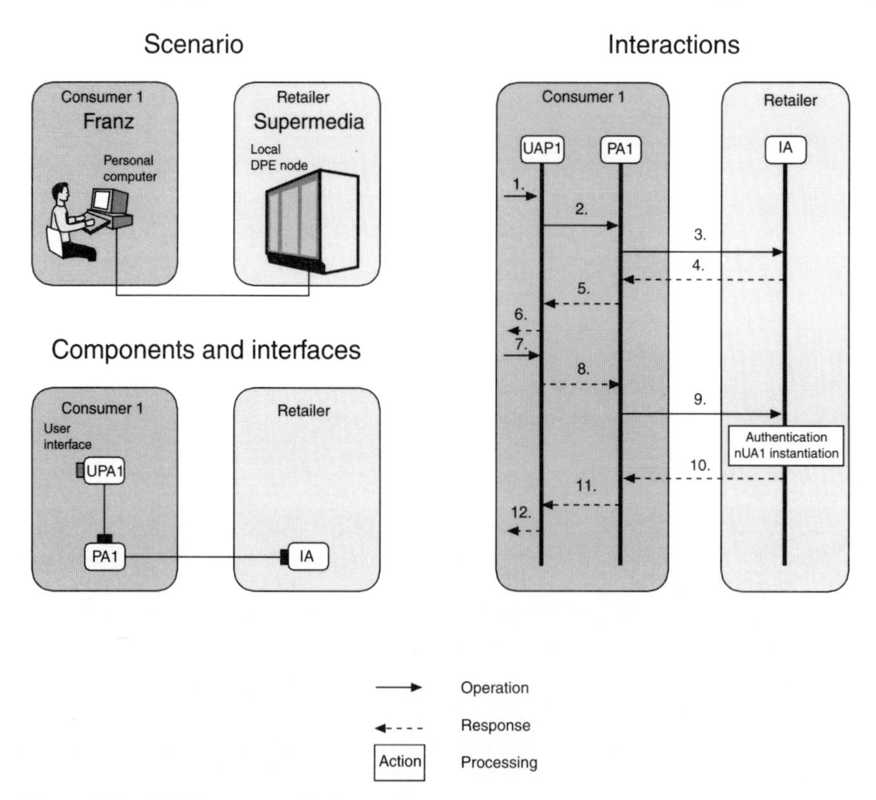

Figure 4.11 Initial contact with the retailer

ational interface of the initial agent (IA) object, whose reference is public.[1] In more detail, the PA triggers a dialing application that sets up an ISDN (over IDSL) control channel between the terminal and the retailer server; the invocation on the interface – which is a DPE-level interaction – can use, for example, the IIOP/TCP/IP protocol stack supported by the ISDN control channel. The counterpart of the consumer terminal, in the retailer domain, is a local (i.e. directly connected to the access loop) DPE node, where some service components are deployed.

The IA replies with a request for authentication information (e.g. username/password) to the PA (4); the PA returns the UAP the indication that authentication information is needed (5), and the UAP prompts the user (6). Note that, depending on the UAP, the authentication information can be given earlier by the human user. However, at the moment of the IA request to the PA, retailers may download forms or any proprietary screen image to collect this information. Also, the default authentication specified by TINA is username/password, but retailers may establish different access authentication mechanisms for their customers. In any case, this information is provided (7–9) to the IA.

The IA verifies the information, and identifies Franz as an existing authorized user for the retailer. Looking into a proper database, the IA determines the interface reference of Franz's user agent (UA) and returns the interface reference to the PA (10). In fact, the component instantiated is of the named user agent (nUA) subtype, since Franz is a recognized user. A message indicating authentication success is given to the UAP and to the user as return value for the operations (11 and 12).

4.7.2.2 Access session interactions

4.7.2.2.1 Service request

At this point an access session is established between Franz and Supermedia. Optionally, a retailer-specific, more sophisticated access-related UAP may be downloaded to the user terminal to replace or enhance the default one.

The overall situation is shown in Figure 4.12. Three TINA components are instantiated: an access-related UAP and a PA in the consumer domain (that is, in Franz's terminal equipment), a nUA in the retailer domain; the IA (in the retailer domain) no longer takes part in the interactions. While it can be foreseen that components such as the IA are deployed close to the access loop (even though this is not "functionally" needed), no assumption is made on the physical location where other components are deployed in the retailer domain: the DPE gives location and distribution transparencies (see section 3.2.1) making the distribution aspects irrelevant, as well as invisible at the service level. For this reason, in Figure 4.12 (access session) the retailer domain is shown as a whole TINA system, with no reference to individual DPE nodes.

At this point, Franz requests the WWW browsing service. Via the user interface of his access-related UAP, he inputs his request (1). The UAP forwards it to the PA in the terminal (2), which sends the request to the nUA in the form of the appropriate operation

[1] The PA can refer to a CORBA name server, which returns the reference to the IA interface of the requested retailer.

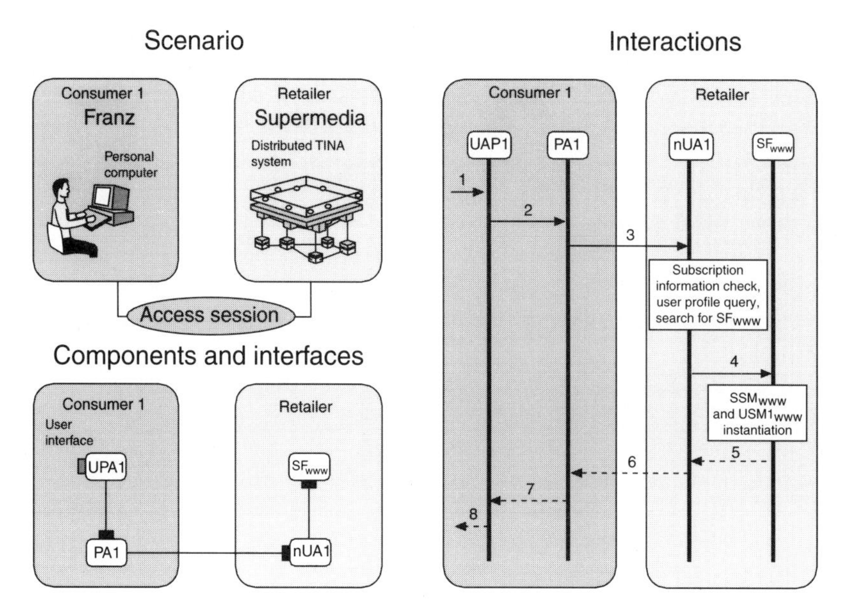

Figure 4.12 Access session

on the nUA interface (3). The PA also provides basic terminal information (terminal type = PC, and so on); the nUA checks whether Franz is authorized to use that service, and determines that he has the required rights. It also queries his user profile to retrieve customization information (user interface options – such as preferred browser, billing records, etc.) and sets up a management context (collection of accounting information, monitoring, etc.) related to the request. Although the Web browsing service "alone" does not require the TINA service architecture, the integration of this service with other services that can be used simultaneously into a common management context and with access services (mobility, reachability, logging, etc.) takes advantage of the TINA approach.

4.7.2.2.2 Service session creation

The nUA searches for the service factory interface of the requested service (WWW browsing) corresponding to Franz's user profile and terminal configuration. It requests the SF to create a service session (4). The SF creates the SSM and USM components and returns to the nUA the interface reference on which the user can access the service (5); this interface is provided by the USM. For this specific service, USM and SSM encapsulate the POP functionality: it is a "TINA view" of a non-TINA service. The functional partitioning between USM and SSM is proprietary and service-specific; the USM may even be absent.

The nUA, using the control channel, returns the successful result of the service request with the reference to the USM interface (6). The WWW browser on Franz's terminals represents the usage-related UAP for the WWW browsing service. Finally, the result of the operations is returned to the corresponding invokers (7 and 8).

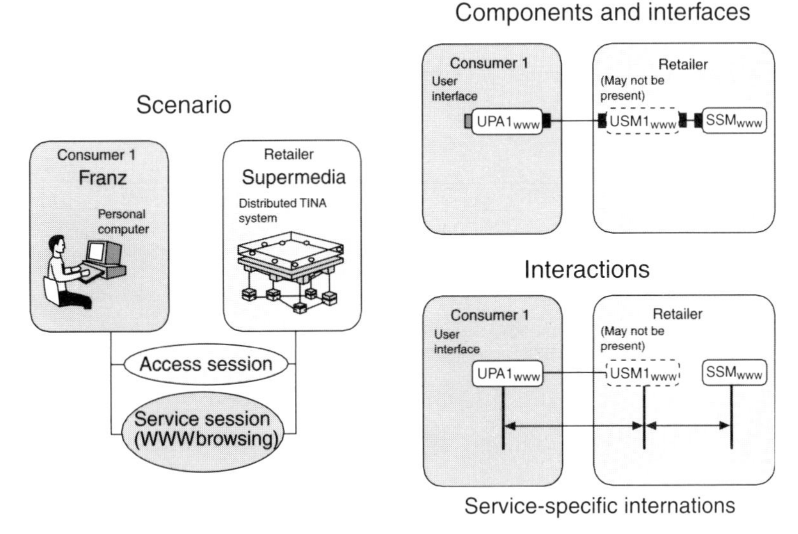

Figure 4.13 WWW browsing service session.

4.7.2.3 Service session interactions

4.7.2.3.1 Usage of the WWW browsing service

At this point, a WWW browsing service session is established between Franz, as a user, and the Supermedia retailer (Figure 4.13). The components involved are the usage-related UAP of the WWW browsing service (i.e. a WWW browser) in the consumer domain and USM and SSM of the WWW browsing service in the retailer domain. The nature of the service implies other stakeholders may be involved, namely the providers of the various Web sites. However, given the nature of the service, interactions between Franz, as a consumer, and the Web servers' domains do not require any TINA mechanism. Hence, in this case, the composed service scenario described in section 4.6.3 does not apply and the other stakeholders are "hidden" by the service logic. This is an example of how the TINA approach may be used in situations where TINA does not need to be "pervasive" in order to add value.

Franz uses the service; he browses the Net searching interesting sites for his purpose. Video clips, images and so on can be provided using the control channel (based on TCP/IP as any Internet browsing). If Franz requires a guaranteed quality of service, more bandwidth is needed. At this point, a TINA-specific solution can be adopted: the service session (i.e. the service logic) may also provide mechanisms to use the IDSL data (stream) channel. Upon a click of a hyperlink, the SSM requests connectivity between a stream interface in the user terminal and another stream interface in a Web server. The TINA connection management, or any other connection management with which the SSM is able to interact, controls this connectivity according to the composed service scenario. For simplicity, we do not consider this option in more detail.

4.7.2.3.2 Request for a multimedia conferencing service

Franz found something interested on the Gulliver Travels Web site: the offer of a cheap vacation in Cuba. He wants to speak to a sales representative, and clicks on a button of the corresponding HTML page. This triggers a request for a multimedia conferencing service. How this is implemented is up to the service designer. Basically, there are two solutions: either the UAP on the terminal contacts the PA[2] or it transmits a HTTP request to the service logic that contacts the UA interface directly (using CGI mechanisms or similar ones).

In both cases, the mechanism to instantiate the new service is similar to that used for the WWW browsing service: the nUA checks subscription information for Franz, searches for the multimedia conference service factory, and requests the creation of a service session. The SF creates the SSM and USM components for the multimedia conference and returns the interface references. What is new with respect to the previous case is that a conferencing service requires a "call" to another user; in our case, the other user is a salesperson, and the information of the called party (the salesperson) address is embedded in the HTML page. In general, the inviting user would have to specify the address of the other participant(s) to the conference. Therefore the service session for the multimedia conference service is established, but there are other actions to perform before the service can be used, which are described below.

The newly created multimedia conferencing service logic (SSM and Franz's USM) retrieves information about the stream interface reference of Franz's microphone, speakers, video camera and monitor, to be used for audio input, audio output, video input and video output respectively. This can be done by querying a database or by issuing the appropriate request to the PA on Franz's terminal.

4.7.2.3.3 Service session invitation

In order to call the other party in the conference, the multimedia conference service logic (either in the UAP in Franz's terminal or "hard-coded" in the retailer domain service components) sends an invitation request to the nUA of the salesperson.[3] We assume for simplicity that this component resides in the domain of Supermedia, but it could also reside in the domain of another retailer. The salesperson is sitting at a multimedia workstation, having both conferencing capabilities and access to the travel agent's Web server (this access is implemented in a proprietary way). The salesperson's nUA in the retailer domain issues an invitation request to Sarah's workstation. This happens by invoking an operation on an interface offered by the PA in the workstation. Sarah, the saleswoman, accepts the invitation.[4]

[2] In this case we have to assume that the service-related UAP (browser) has the PA interface reference.

[3] The nUA could be a specific to a role ("Gulliver Travels salesperson"), and not the specific individual (Sarah) that serves the request. This feature is part of the added-value access services that a retailer may provide to its customers.

[4] In this situation, we can assume that the acceptance is "automatic". In general, the acceptance could require a decision from the human user; in this case the invitation is delivered, then the interaction terminates; at any time (provided the session is still active) the PA can send a mesage indicating invitation acceptance.

Upon acceptance of the invitation by the called party (in our case, Sarah), a multimedia conference USM component is created for that party. The information about stream interfaces for the salesperson (audio and video input/output interface references) is retrieved.

4.7.2.3.4 Stream binding establishment

Although the involved parties in the conference are just two, the service is such that other parties may be added in the future. Also, there is the need to combine audio and video inputs from all parties into a single audio and video output (e.g. the combination of the views of each party and the "sum" of the voices). For this, an audio/video bridge has to be involved in the service session. This is a special resource: it corresponds to the resource role in the service session graph and to a service support component. This component offers a stream interface for audio and video flows and an operational interface to the SSM for control. These interfaces have to be instantiated (or made available) within this particular service session. We assume that this service also supports data stream communication. Video clips or still images can be sent from a multimedia server (also acting as service support component) to the conference participants and shown at the multimedia conference. This is also achieved through the bridge logic: inputs come from stream interfaces corresponding to each party, plus any multimedia server; they are merged into a "unified output" and sent to each party. There is a stream flow for each of these media; this flows from a *source* stream flow end-point to a *sink* stream flow end-point.

At this point, everything is ready for the establishment of the stream binding between Franz (party), Sarah (party), the bridge (resource) and – if required – any multimedia server (resource). For this, the SSM contacts the connection management giving information on stream interfaces as input parameters. The connection management sets up the necessary connectivity: from each party to the bridge, and from the bridge to each party (point-to-multipoint). Chapter 5 shows the stream binding implementation using the TINA connection management architecture.

4.7.2.3.5 Usage of the multimedia conferencing service

Audio, video and multimedia data information is now flowing from Franz's terminal, via the bridge, to Sarah. If any service-specific software is required in the users' terminals, this may be downloaded at any time upon the initiative of the service logic (if foreseen by the service designer). In our case, this could be software to control video clips (similar to video-on-demand service) or any other feature the retailer may wish to provide. Franz and Sarah discuss matters; Franz views video clips and still images of wonderful tropical beaches and historical monuments ...

4.7.2.3.6 Invitation to other parties

Franz decides to invite three other parties to the conference: his friends Hiroshi, Barbara and Tanja. He requests the invitation to the usage-related UAP in his terminal, giving the identifier of these users. The UAP sends the invitation request to Franz's USM, which checks whether Franz is authorized to invite new parties to that particular service session. Assuming that he is, the request succeeds and the USM forwards the request to the SSM.

The SSM sends an invitation to each invited user's nUA. The nUAs in the retailer domain issue an invitation request to each user's terminal PA, in the same way for the salesperson. Note that, in this case, we cannot assume that each person is sitting at their terminal waiting for a call: the nUA has to select the right PA interface reference (a user could be "associated" to many terminals depending on his or her physical location, type of call, and so on) and, if available, send the invitation message to it; this causes an event equivalent to the telephone ringing in a POTS call. Depending on what happens (busy, no answer, etc.) the nUA has to take a decision following a certain policy, typically dependent on the user profile: store the invitation, discard it, queue it, etc.

For this example, we assume that each user is available, receives the invitation and accepts it. Correspondingly, USM components are created for each new party, information on stream interfaces are retrieved and the stream binding is modified to include the new parties; connectivity is established between the new parties and the bridge. Multimedia flows are shared by all parties, included the newly invited ones.

The four customers and the travel agent discuss the offer, view video clips and images as before. At the moment of taking the decision, Franz requests that the salesperson is momentarily excluded from the session: the domain user service session corresponding to her is suspended. They decide among themselves to take that offer, and re-establish the conference with Sarah; her domain user service session is resumed.

4.7.2.3.7 Electronic payment

The four friends now have to pay for their vacation. We assume that they are all subscribers to an electronic payment service with the retailer. The service consists of the fact that subscribers (properly authenticated) can ask the retailer to pay another subscriber a certain amount; that amount will be charged to the payer and credited to the payee in the unified telecommunications services bill, sent monthly by the retailer to its subscribers. Franz, Hiroshi, Barbara and Tanja create a service session for this service, involving each user, the retailer and the travel agent (i.e. the salesperson Sarah).

At this moment, each "character" of our story has now an access session between themselves and the retailer. Franz, Hiroshi, Barbara and Tanja have each, individually, a service session between themselves, the retailer and the travel agent for the electronic payment service; Sarah has four electronic payment service sessions with each payer and the retailer. Furthermore, the four friends and the salesperson have *jointly* a multimedia conference service session with the retailer.

The four friends and the salesperson, individually, undergo service-specific "strong" authentication for the electronic payment. Then, each of them authorizes a $800 (the price of the Cuba vacation) payment to Gulliver Travels. The electronic payment service logic (SSM and USMs) logs the transaction, forwards the transaction information to the billing systems to charge $800 to each of the four friends and credit 4*$800 to the travel agent. Then each person terminates their own electronic payment service session.

4.7.2.4 Closing

All customers say goodbye to Sarah, who then quits the multimedia conference service session. The domain user service session associated to her terminates. The corresponding USM component in the retailer domain is deleted.

Our four friends stay on-line for some moments more, chatting about their vacation. At a certain time they decide to take leave of each other. At this point, Franz terminates the service session. All domain user service sessions terminate consequently. All USMs, and then the SSM, for the multimedia conferencing service are deleted. Each person individually terminates their own access session. The nUA interface for that access session is deleted, or in any case made unavailable.

Have a nice vacation, Franz, Hiroshi, Barbara and Tanja!

4.8 WRAPPING UP

This example shows how the TINA service architecture actually works in real situations where TINA-specific services (such as a multimedia conference with guaranteed bandwidth) and other services (such as World Wide Web) coexist and give added value to each other. The TINA retailer is the "system integrator" of a variety of service and transport platforms, and the service architecture is the conceptual tool by which this is achieved. The TINA session concept – in particular the separation between access, service and communication session – is the enabling factor. This topic is expanded in Chapter 7, where the integration of TINA and non-TINA systems is examined in detail.

Chapter

5 Network resource architecture

5.1 INTRODUCTION

The TINA network resource architecture (NRA) is the TINA component architecture that specifies concepts and principles for the specification and design of functions that control and manage a TINA network. Figure 5.1 illustrates the scope of NRA in terms of the TINA business model and inter-domain reference points.

The scope of NRA spans two classes of functions:

- Those that are traditionally viewed as network control functions, i.e. that provide connection setup and control, including signaling, route selection, and bandwidth management.
- Those that are traditionally viewed as management functions, i.e. that deal with network resource management in support of connection setup and control. The management functions deal with fault management, configuration management, accounting management, performance management, and security management. These management functional areas are together referred to as FCAPS.

In previous generations of network architectures (until the late 1980s) these two classes of functions were treated separately, each with its own architectural principles. Thus, for example, network control functions are usually based on a distributed architecture and the functions reside in each network element. Interoperability of the control functions across administrative domains (network operator jurisdictions) is viewed as of paramount import-

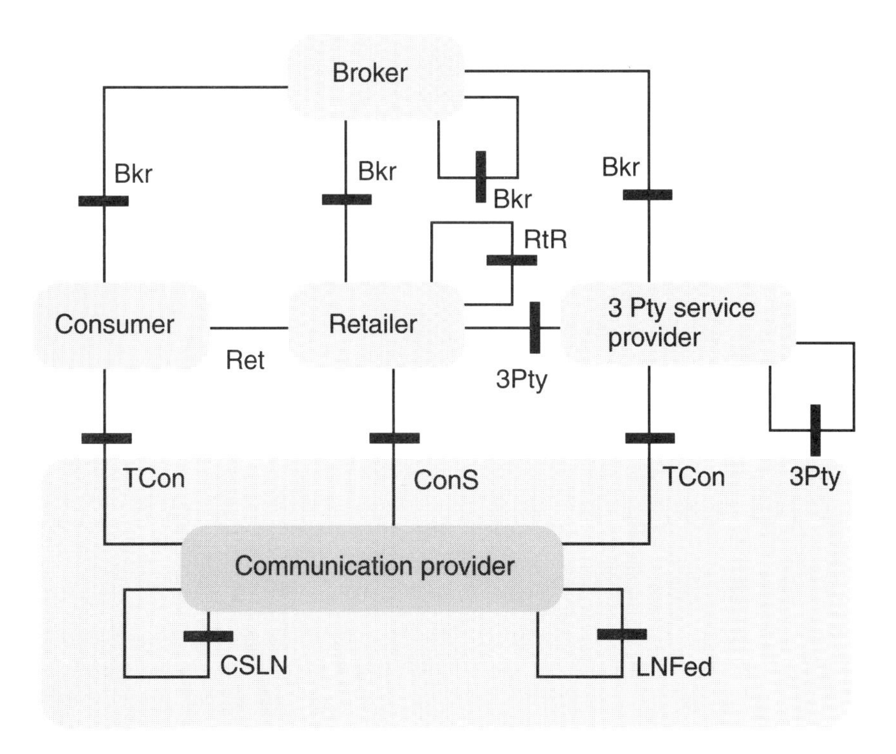

Figure 5.1 Scope of TINA network resource architecture

ance and is ensured through the usage of standard signaling protocols. On the other hand, network management functions are less distributed, and functions are often centralized within each network operator domain. Interoperability between management functions (operations support systems) in different administrative domains is largely absent, and if present, is realized using pairwise business agreements and proprietary protocols. The rationale for distinguishing the two classes in such a manner is that network control functions need to operate in a real-time environment while the management functions do not have any such real-time constraints. Reinforcing this separation, the two classes of functions are deployed using different computing platforms. While network control software is deployed on special-purpose real-time platforms, management software is typically used on general-purpose computing platforms.

A basic premise in the TINA architecture, more specifically in the NRA, is that the above rigid separation between network control functions and network management functions should be removed and both classes of functions should be designed based on a common computing architecture. Use of such a common computing architecture will result in many benefits: it leads to uniformity in design; interactions between control and management functions is facilitated; sharing and exchange of information between the two classes of functions do not require any special mechanisms; both classes of functions can be designed with distribution in mind and a function is distributed and/or replicated taking into account its scalability and performance requirements; re-use of platforms for both classes of functions is enabled leading to a reduction in procurement costs; and both classes of functions can be resident in a computing node. Real-time requirements of control functions as well as some management functions (such as alarm surveillance and customer network management) can be handled in a uniform manner using engineering techniques that do not alter the basic computing architecture.

Towards this unification of network control and management functions, NRA defines the architecture of TINA network control and management functions using a common set of architecture concepts and principles.[1] In addition to this unification, NRA has been developed with the following objectives:

- NRA should define an end-to-end connection service that transports information between TINA applications. This connection service should be based on the concept of stream flows and stream interfaces defined in the TINA computational model.
- The target applications in TINA require complex multimedia and multipoint communication sessions. The NRA connection service will support setup and control of such communication sessions.
- In providing this connectivity service to applications, NRA should ignore applications network technology-dependent aspects. Applications should be able to specify bandwidth and other quality of service (QoS) attributes in a technology-independent manner. NRA should translate these requirements to technology-specific parameters and setup connections accordingly.
- It is possible for the network to consist of parts based on different network technologies (such as ATM, frame relay, N-ISDN). Interworking between these different technology sectors is a concern of the NRA and interworking details should be transparent to applications. Further, the network may span multiple administrative domains. NRA should ignore application details of interworking across such domain boundaries.
- The focus of NRA is on network level aspects. The NRA concepts should complement concepts and models developed in other standards bodies and forums (such as ATM Forum and ITU-T) for technology-dependent and network element level control and management [ATMF95, ATMF97, ITUT92b, ITUT96, ITUT92c].

5.1.1 NRA and related standards

Although the NRA is distinguished from other network management and control architectures in terms of its objectives and scope, the relationship between NRA and these other architectures is complementary and synergistic. In many ways, the NRA can be viewed as a consolidation of the principles embodied in other architectures in one coherent framework.

[1] The term "network resource architecture" signifies this unification.

In the realm of telecommunications network management, the ITU-T standard telecommunications management network (TMN) as defined in ITU-T Recommendation M.3010 sets the basic foundation for interoperable management system specification and design [ITUT92c]. Some of the key principles of the TMN architecture, such as management of network resources using abstract representations of the resources (called managed objects) and decomposition of management functions into several layers, are embodied in the NRA. An aspect that distinguishes the NRA (and more generally TINA) from TMN is that the NRA is based on a distributed computing architecture that relies on a DPE that provides distribution transparencies to applications. This makes possible a choice of different communication protocols between management applications in different system configurations as well as flexible placement of management applications in computing systems.

Currently, there is work in several technology forums, such as the ATM Forum and the Sonet Interoperability Forum, on the development of technology-specific management information models based on the TMN principle of separating network management functions into several layers: *element layer* (EL), *element management layer* (EML), and *network management layer* (NML) [ITUT96, ITUT92c]. This work is based on the generic functional architecture of transport networks defined in ITU-T Recommendation G.805 [ITUT95]. The NRA is also based on G.805, and so many NRA concepts have their counterparts in the ATM and Sonet management information models. The NRA is distinguished from these by the fact that the focus of NRA is only on technology-independent network level management aspects, and it is also concerned with management of stream flows. Given the similarity of NRA concepts to the ATM and SDH/Sonet technology models, a network operator may choose to use these models to implement technology-specific management systems, and implement the NRA on top of these systems.

In traditional network control, network connections are established and released using signaling protocols that specify user-to-network and network-to-network interactions for connection management. Signaling standards have been defined in intelligent network (IN) and BISDN standards. As mentioned earlier in this section, the NRA does not define a separate signaling architecture but incorporates connection management as an aspect of network resource management and defines the architecture for all management functions based on a common computing architecture. This approach enables use of distributed computing techniques and platforms for connection management as well as other management functional areas in a uniform manner.

5.1.2 Organization of the chapter

The objective of this chapter is to present an overview of NRA based on the descriptions contained in NRA-related TINA documents available as of April 1997 [TC-AN97, TC-CN94, TC-CI97, TC-MI96] This chapter is organized as follows.

Section 5.2 describes the overall structure of a TINA network. This description is based on the TINA network resource information model (NRIM) which specifies the information elements for representing the topological and connectivity structure of a TINA network [TC-AN97, TC-CI97]. Each of sections 5.3 through 5.6 describes the architecture of a specific management functional area. As of April 1997, only the following management functional areas have been addressed in NRA: connection management, topology configuration management, fault management, and accounting management. The architecture of each of these management areas is described in a separate paragraph. Section 5.7 presents a preliminary view on how management of internet can be incorporated into NRA. Section 5.8 concludes this chapter by highlighting the distinguishing aspects of NRA.

5.2 TINA NETWORK RESOURCE INFORMATION MODEL

The first step towards the design of control and management functions for a network is the development of an abstract model of the resources to be managed. In TINA, such a model, called the *network resource information model* (NRIM), has been developed. This section presents an overview of the NRIM. The NRIM is a pure resource model and is independent of the computational (i.e. functional) architecture of the resource management functions. In this sense, the NRIM is a "common information model" for all the resource management functions. It is a technology-independent information model. A detailed description of the NRIM is presented in [TC-AN97, TC-CI97].

5.2.1 Overall structure of a TINA network

A network modeled by the NRIM, hereafter referred to as a *TINA network*, is a transport network that is capable of transporting multimedia information. The information traffic carried by the network will be heterogeneous in terms of data formats, bandwidth requirements, and other QoS characteristics. The network traffic in a TINA network can consist

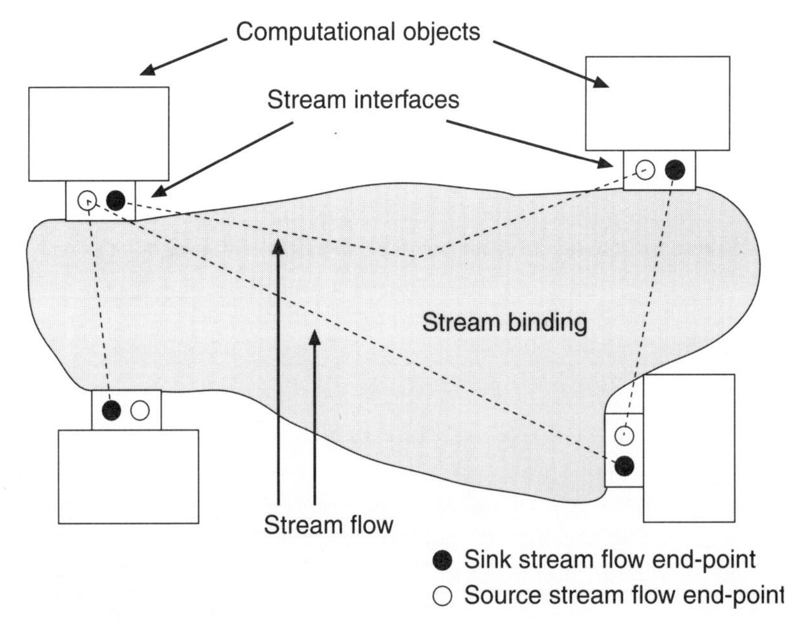

Figure 5.2 TINA network (logical view)

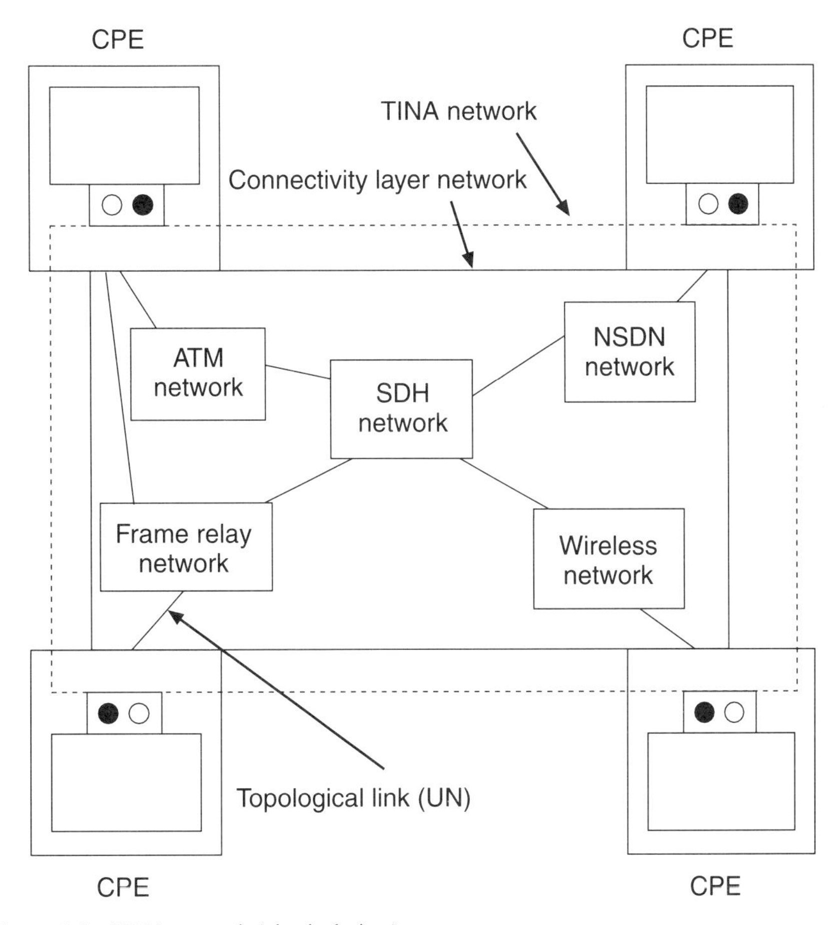

Figure 5.3 TINA network (physical view)

of interrelated multimedia streams and the TINA network transports such traffic ensuring stream synchronization. The application level end-points (TINA logical view) in the TINA network model are stream interfaces associated with TINA applications or multimedia devices (see Figure 5.2).

Connectivity resources at this level are called *stream flow connections* (SFC). A stream flow connection is bounded by two or more *stream flow end-points* (SFEP). An SFEP is either an information source or a sink, but not both (i.e. SFEPs are unidirectional). A source SFEP can be bound to one or more sink SFEPs (providing point-to-multipoint connectivity). *Stream flow end-point pool* (SFEP pool) is a modeling construct that aggregates SFEPs belonging to an application or a multimedia device. (A SFEP pool is the representation of a stream interface from a resource perspective.) A *stream binding* is a modeling concept that represents a collection of stream flow connections that have been grouped together for some purpose at the application level. An SFEP can terminate only one stream flow. Associated with a SFEP is the characteristic information accepted/delivered at that SFEP. These properties include frame structure identification, QoS, etc. The frame structure and QoS of the source and sink stream flow end-points bound by a stream flow connection need not be identical.

In the physical view, a TINA network is divided into two main components: one is the *connectivity layer network* and the other is formed by the communication resources contained in *customer premises equipment* (CPE) (see Figure 5.3). A CPE may be either a simple terminal device (telephone or multimedia device) or a computing system in which TINA-compliant applications are deployed. A connectivity layer network is a transport network consisting of a heterogeneous collection of switching resources, transmission resources, and adapters. A connectivity layer network is made up of components of different technologies, such as ATM, frame relay, narrowband ISDN, wireless, SDH, or PDH, and is capable of transporting different types of information (Figure 5.3 shows some of the possible technologies). Note that it is possible that a CPE can be attached to several such networks.

A communication end-point at which a connectivity layer network accepts or delivers information is called a *network flow end-point* (NFEP) (see Figure 5.4).[2] From the perspective of a connectivity layer network, an NFEP may be either a source, sink, or both (contrast this with an SFEP). Associated with an NFEP is a characteristic information accepted/delivered at the NFEP. These properties include frame structure identification,

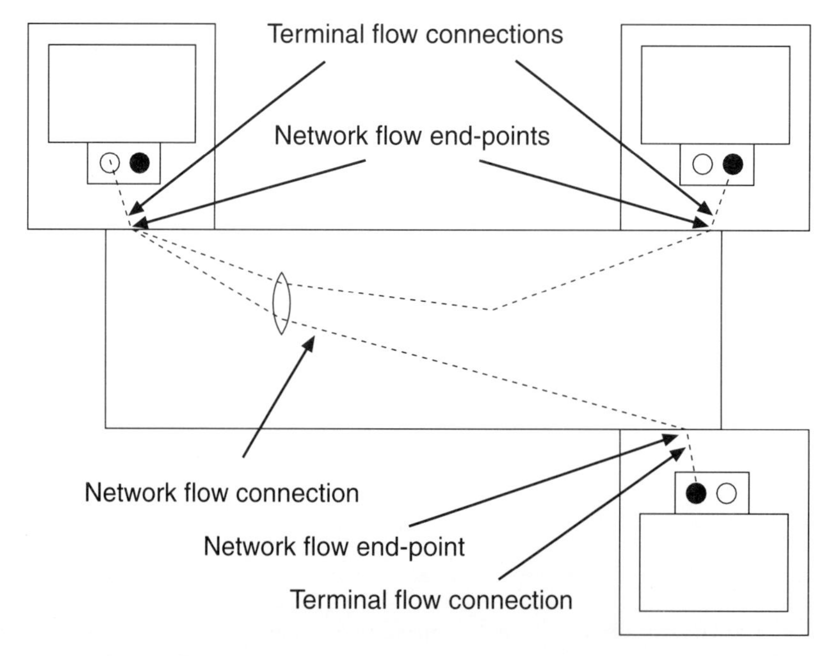

Figure 5.4 Stream flow, network flow and terminal flow connections

[2]A NFEP is a technology-independent representation of a network trail termination point, which is an access point to a layer network

QoS, etc. A *network flow connection* (NFC) is a connectivity resource that transports information between a group of NFEPs. An NFC has one of the following configurations:

- A point-to-point bidirectional connection between two NFEPs
- A point-to-point unidirectional connection between two NFEPs. One of the end-points is designated as the *root* NFEP, and the other end-point is designated as the *leaf* NFEP. Information is transported from the root NFEP to the leaf NFEP.
- A point-to-multipoint unidirectional connection between two or more NFEPs. One of them is designated as the root NFEP and the others are designated as the leaf NFEPs. Information is transported from the root NFEP to the leaf NFEPs.

A stream flow connection is composed of one or more network flow connections and two or more *terminal flow connections* (TFCs) (see Figure 5.4). A terminal flow connection is a connectivity resource that transports information either from an SFEP to an NFEP or vice versa, or to another SFEP within the same CPE. It is possible that the frame structure and QoS associated with the SFEP and NFEP bound by a terminal flow connection are different, in which case the TFC performs the necessary adaptation. It is possible that multiple TFCs have the same NFEP; i.e. several stream flows can be multiplexed over a single network flow connection.

5.2.2 Layer network and trail

In networking literature, the concept of *layer network* (LN) is used to denote a network that is based on a single technology and that transports information of a specific format, referred to as the *characteristic information* of the layer network. Examples of layer networks are: ATM virtual path (VP) network, ATM virtual channel (VC) network, SDH VC4 path network, and frame relay network. The concept of a layer network was originally defined in the ITU-T Recommendation G.803 which describes the functional model for SDH transport networks [ITUT92a]. This concept has been adopted in the subsequent ITU-T Recommendation G.805 [ITUT95] that describes the functional and structural architecture of transport networks in a generic manner.

A connectivity layer network is made up of one or more layer networks. A layer network may be related to another in a connectivity layer network in one of two ways:

- *Peer-to-peer relationship*: This relationship exists when information delivered by one layer network is adapted and given as input to the other and vice versa. This relationship is symmetric, and is referred to as a layer interworking relationship. An example of this relationship is adaptation of frame relay to ATM (VP/VC), and vice versa.
- *Client–server relationship*: This is described later in section 5.2.4. This relationship exists when a group of link connections in one layer (the client layer) network is served by a trail in the other layer (the server layer) network. Such a

group of link connections is called as a *topological link*. See section 5.2.4 for some examples.

The network resource that transports information across a layer network between two or more end-points in the layer network is called a *trail*. Thus, a trail is defined relative to a layer network. For example, a virtual path (VP) trail is a resource that transports ATM VP cells across a VP layer network. A trail may have one of the following configurations: point-to-point bidirectional, point-to-point unidirectional, or point-to-multipoint unidirectional. The end-points of a trail are called *network trail termination points* (NWTTPs). Associated with each NWTTP is bandwidth and QoS information (for each direction of traffic), and some additional information that may vary with each technology. For example, in an ATM VP layer network, associated with a NWTTP is a virtual path identifier (VPI), that is used as the local reference for the ATM VP trail.

5.2.3 Structure of a layer network

5.2.3.1 Subnetworks and links

A layer network is decomposed into *subnetworks* that are interconnected by *links* between them (see Figure 5.5). A subnetwork is an interconnected group of network elements. Connections can be established and released dynamically between termination points (called edges, see below) on the boundary of the subnetwork. A link represents a topological connectivity between either two subnetworks or a subnetwork and a CPE. Each subnetwork may be further decomposed into smaller subnetworks interconnected by links until the subnetwork is equivalent to a single network element (switch or digital cross-connect). A link is configured using one or more trails in an underlying server layer network. To distinguish between the general concept of a trail, and the specific use of a trail for configuring a link, the concept of topological link is defined. A topological link is a logical or physical transmission path interconnecting two network elements and is supported by a trail in a server layer trail network. For example, an SDH path interconnecting two ATM NEs is represented as a topological link in the ATM VP layer network. The bandwidth of a topological link is determined by the bandwidth of the underlying server trail. The end-points of a topological link, called *topological link termination points* (TLTPs), are the points at which adaptation of client layer information to server layer information occurs. Thus, a topological link is configured using exactly one trail in the underlying server layer network, and a link is configured using one or more topological links. See Figure 5.6.

Figure 5.6 shows a layer network configuration consisting of two subnetworks. The subnetworks are interconnected by two topological links. A link has been configured using the two topological links. This link represents the aggregate capacity for connectivity between the two subnetworks. Each termination point of a link is called a *link termination point* (LTP).

A link can be configured using topological links in any of the following ways:

- 1:1 configuration: A link is configured using one topological link by assigning the entire bandwidth of the topological link (server layer trail) to the link.

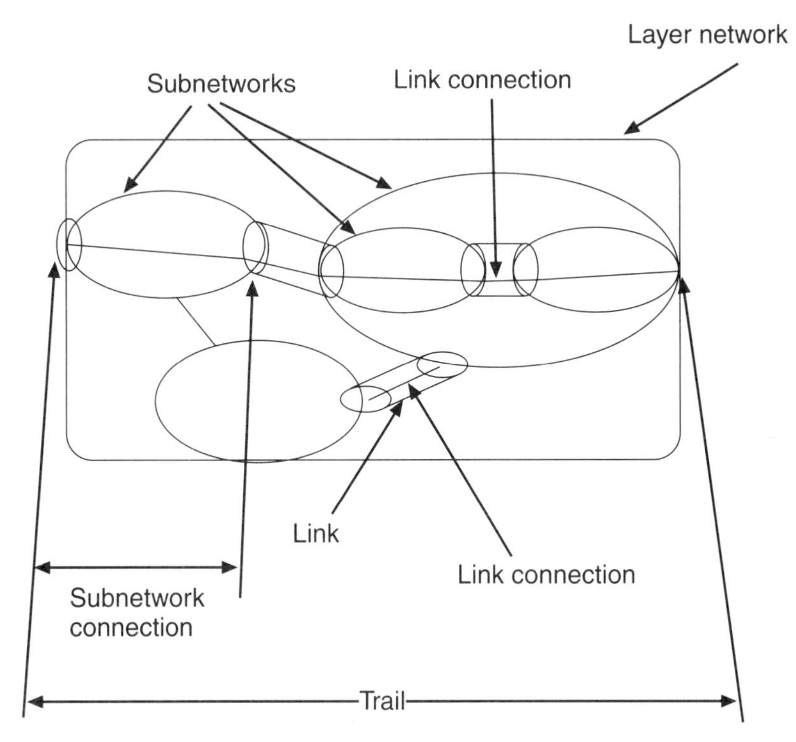

Figure 5.5 Structure of a layer network

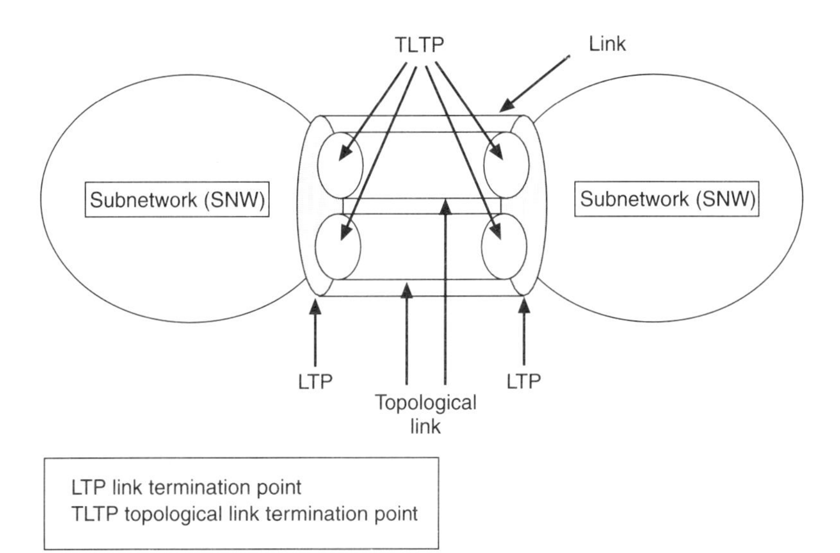

Figure 5.6 Links and topological links

- N:1 configuration: A link is configured using one topological link by assigning to the link only a portion of the bandwidth of the topological link. Many such links can be configured using a topological link.
- 1:N configuration: A link is configured using a set of topological links by assigning the entire bandwidth of all topological links in the set. The topological links may be terminating on different network elements. This is the case shown in Figure 5.6.

Another possibility is to configure a link as an aggregation of two or more links. Such links are sometimes called composite links. These are very useful in representing aggregate capacity of the interconnections between two subnetworks, such as, for example, between two subnetworks in two adjacent administrative domains.

Irrespective of how a link is configured, the network resource that transports information across a link is called a *link connection*. At link configuration time, the link connections associated with the link may or may not be established. This is a technology-dependent matter. In ATM, a link represents the potential for link connections, and link connections are not created at link configuration time. In SDH, a link represents a bundle of link connections, and in this case, all link connections associated with the link are created at link configuration time.

The network resource that transports information across a subnetwork between two or more end-points in the subnetwork is called a *subnetwork connection* (see Figure 5.5). Reflecting the partitioning of a layer network into subnetworks and links, a trail is made up of one or more subnetwork connections and link connections. Similarly, reflecting the partitioning of a subnetwork into subnetworks and links, a subnetwork connection may also be made up of one or more subnetwork connections and link connections (see Figure 5.5).

5.2.4 Client–server relationship

As mentioned earlier, two layer networks may have a client–server relationship. This relationship exists when the transport capabilities of one layer (the server layer) are used in the other layer (the client layer). More specifically, this relationship is established when a topological link in the client layer network is configured using a trail in the server layer network (see Figure 5.7). Some examples of the client–server relationship between layer networks are listed below:

- A topological link in an ATM VP layer network is configured using a path (the name of the trail in the path layer) in the SDH path layer network.
- A topological link in an ATM VC layer network is configured using a trail in a VP layer network.
- Link connections in a SDH path layer network are provided by a paragraph (name of the trail) in the transmission media layer network.

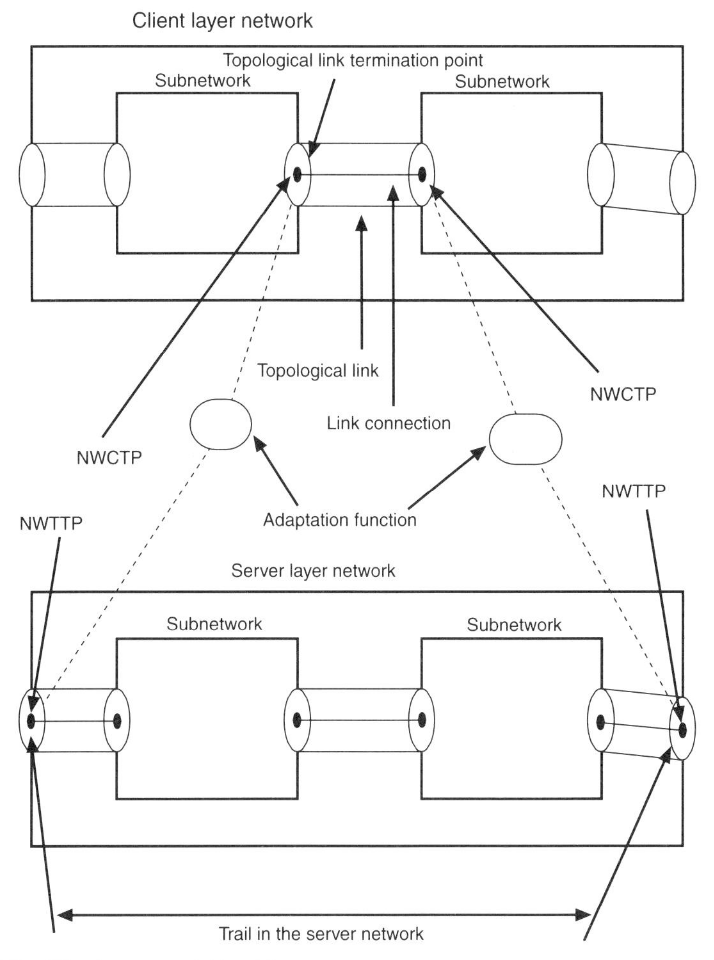

Figure 5.7 Client–server relationship

The end-points of a link connection are called *network connection termination points* (NWCTPs). It should be noted that while a trail or a subnetwork connection may be either a point-to-point or a point-to-multipoint connection, a link connection is always a point-to-point connection.

5.2.5 Edges and network connection termination points

As described in section 5.2.3, it is possible that a subnetwork is partitioned into two or more subnetworks interconnected by links. Reflecting this partitioning, a subnetwork connection may also be made up of two or more subnetwork connections and link connections. In general, the lifetime of a subnetwork connection and its component subnetwork connections and link connections may be different. That is, when a subnetwork connection is deleted, some of the component subnetwork connections and link connections may continue to exist. Similarly, a subnetwork connection may be set up using existing subnetwork connections and link connections. These capabilities are useful for rerouting trails and subnetwork connections upon failures. To allow for this generality, the NRA distinguishes between an end-point of a subnetwork connection and an end-point of a link connection. This distinction is especially important for unidirectional subnetwork connections and link connections. The two kinds of end-points are distinguished using the concepts of network connection termination points and edges as described below (see Figure 5.8):

- *Network connection termination point* (NWCTP): A termination of a link connection is called a network connection termination point. A link connection is only a point-to-point connection, and thus a link connection has only two network connection termination points.
- *Edge*: An extremity of a subnetwork connection is called an edge.[3] A point-to-point subnetwork connection has two edges, and a multipoint subnetwork connection has more than two edges. An edge of a subnetwork connection is bound to a network connection termination point. This binding may change during the lifetime of the subnetwork connection.

Reflecting the partitioning levels of a subnetwork, a subnetwork connection in a composite subnetwork is partitioned into subnetwork connections in the component subnetworks. In such a situation, an edge of the composite subnetwork connection and an edge of the component subnetwork connection will be bound to the same NWCTP (see Figure 5.8).

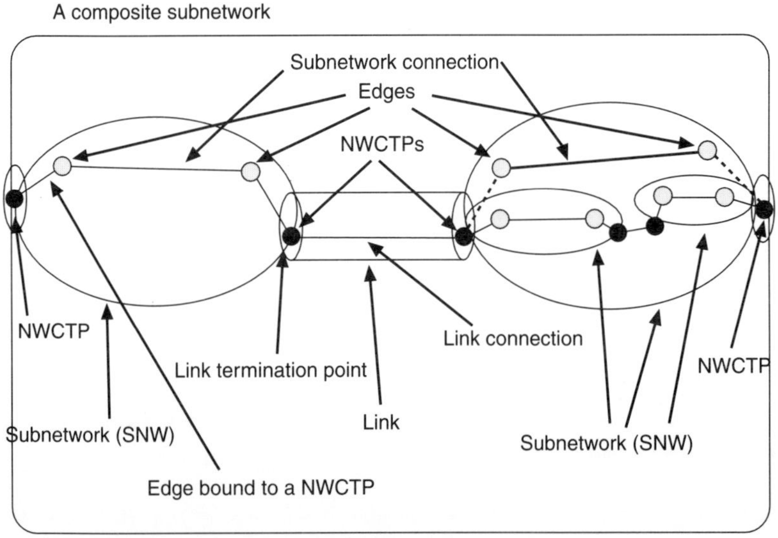

A composite subnetwork
Subnetwork connection
Edges
NWCTPs
NWCTP
Link termination point
Link connection
NWCTP
Subnetwork (SNW)
Link
Subnetwork (SNW)
Edge bound to a NWCTP

Figure 5.8 Relationship between edges and NWCTPs

[3] The concept of edge is identical to the concept of subnetwork termination point defined in ITU-TG.85x Recommendations [ITUT96] and ATM Forum Network View MIB [ATMF97]. The term 'edge' is used here for historical reasons.

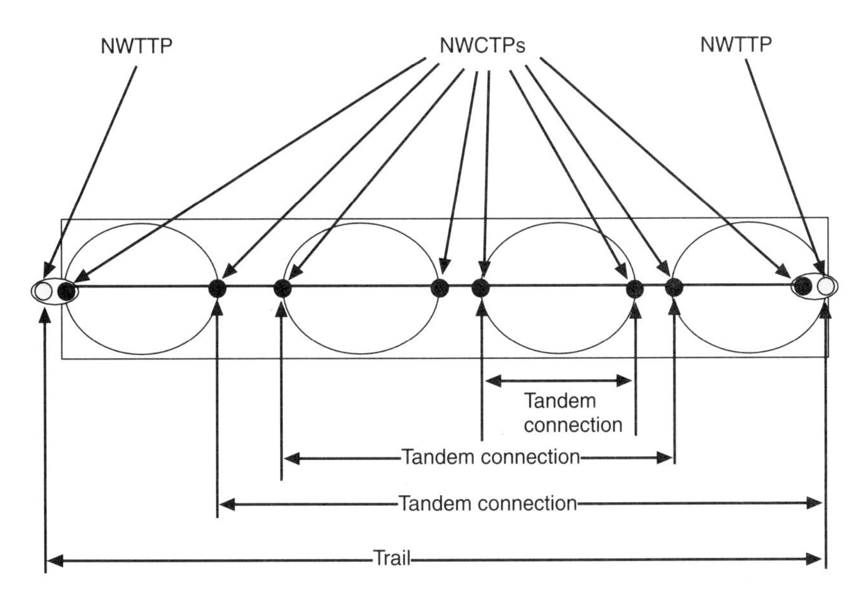

Figure 5.9 Tandem connection

5.2.6 Tandem connection

A tandem connection is an arbitrary series of contiguous subnetwork connections and/or link connections and is used to represent an arbitrary segment of a trail. The extremities of a tandem connection are either network connection termination points or trail termination points. As in the case of a trail or a subnetwork connection, a tandem connection may be either point-to-point or point-to-multipoint.

Figure 5.9 illustrates tandem connections. As can be seen from the figure, a limiting case of a tandem connection is a subnetwork connection. The concept of tandem connection is very useful in situations where a trail spans multiple subnetworks and the subnetworks are under the control of different administrations.

5.2.7 TINA network from the perspective of a connectivity provider

The modeling concepts described thus far are useful for describing a TINA network from a global perspective without any regard to administrative domain boundaries. Such a global view is useful for introducing the various concepts, but it is not the appropriate base on which network control and management functions can be designed. For that purpose, what is needed is the *view of a TINA network from the perspective of a connectivity provider*.

Typically, different portions of a TINA network will be under the control of different network administrations, and the structure of the TINA network seen by each administration may differ. This is a necessary consequence of differing business arrangements between different connectivity providers.

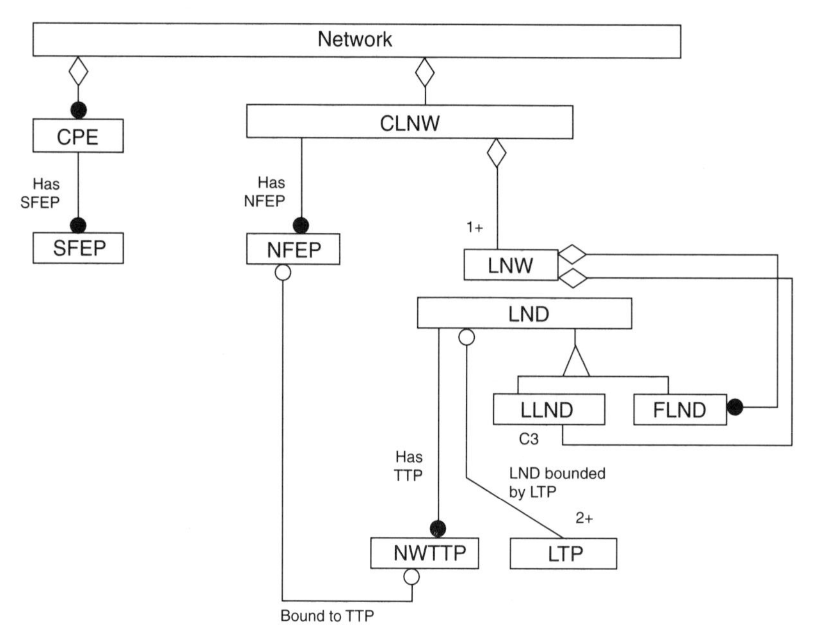

Figure 5.10 Topology view in a connectivity provider domain: 1

From the perspective of a connectivity provider, a TINA network is seen as being composed of a connectivity layer network to which CPEs are attached. The connectivity layer network contains a number of layer networks and access points to them. Different portions of a layer network may be under the control of different network administrations. From the perspective of a connectivity provider, a layer network is seen as being made up of one or more *layer network domains* (LNDs). The portion of a layer network that is controlled by the connectivity provider is called the *local layer network domain* (LLND), and each layer network portion that is under the control of another connectivity provider is called a *foreign layer network domain* (FLND). A LND (local or foreign) consists of a subnetwork (referred to as the top-level subnetwork) on which one or more links terminate. Each such link interconnects the top-level subnetwork of the LND with either the top-level subnetwork of another LND or a CPE. A connectivity provider does not see any further decomposition of the top-level subnetwork of a foreign LND. Only the decomposition of the top-level subnetworks of the local LNDs is visible to a connectivity provider. Figures 5.10 and 5.11 illustrate through OMT diagrams the topology view in a connectivity provider domain.

Similarly, the view of a trail that spans multiple layer network domains is different in each domain. The view held by a domain depends on whether the domain originated the trail, i.e. initiated the trail setup. The originator domain has an end-to-end view of the trail; i.e. it sees the trail as being made up of one tandem connection in its domain, and a tandem connection in each foreign layer network domain that it requested to be set up as part of the trail setup. A non-originator domain sees only the tandem connection in its domain,

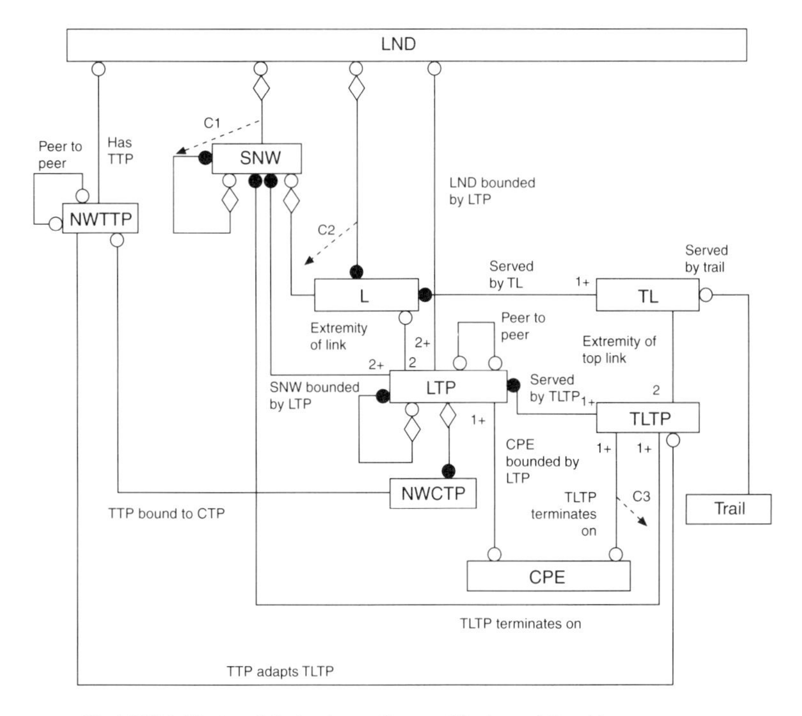

C1: A SNW object participates in exactly one of the two relationships
C2: A L object participates in exactly one of the two relationships
C3: A TLTP object participates in exactly one of the two relationships

Figure 5.11 Topology view in a connectivity provider domain: 2

and does not have an end-to-end view of the trail. Further, a connectivity provider has a detailed view of all components of the tandem connection in its domain (i.e. the subnetwork connections and the link connections that make up the tandem connection), but has only the abstracted end-to-end view of a tandem connection in a foreign domain. A similar asymmetry in views held by domains occurs also in a situation where a tandem connection spans multiple administrative domains. Figure 5.12 shows the OMT diagram for the connectivity view in a layer network domain.

5.3 CONNECTION MANAGEMENT ARCHITECTURE

5.3.1 Levels of connection management

The scope of connection management in TINA is very broad. It ranges from the management of communication sessions and stream flow connections (the highest level) to management of cross-connections within a network element (the lowest level). Figure 5.13 illustrates the various levels of connection management in TINA including the computational objects associated with each connection management level. A brief description of each connection management level (starting from the highest) is given below:

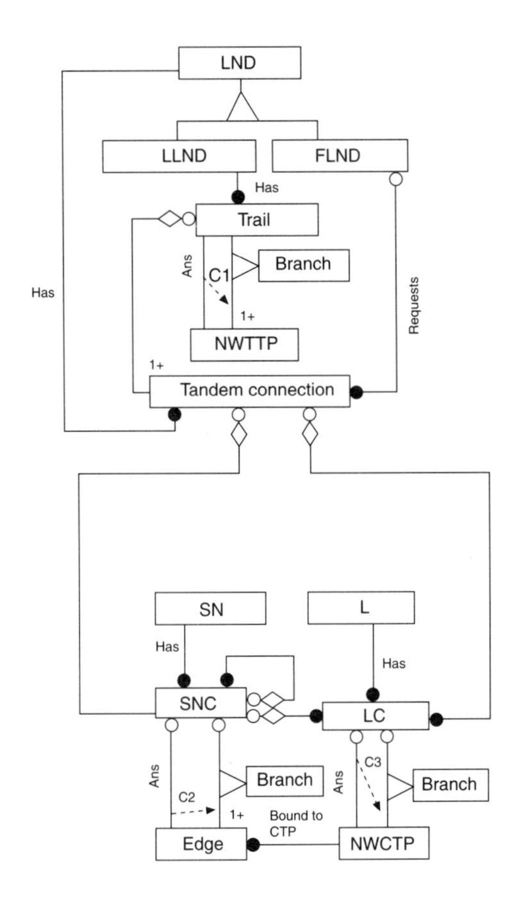

C1: A NWTTP object participates in exactly one of the two relationships
C2: An Edge object participates in exactly one of the two relationships
C3: A NWCTP object participates in exactly one of the two relationships

Figure 5.12 Connectivity view in a layer network domain

- *Communication session level*: The connection management function provided at this level is setup and management of stream flow connections (which are the end-to-end application level connections). Service level components are the clients of this level. This functionality is provided using the concept of a *communication session*. A communication session is an environment (or context) for establishing and managing a number of stream flow connections. It represents together an aggregate of stream flow connections, the components (computational objects) that manage the stream flow connections, and policies (and profiles) governing this management.[4] The client of a communication session is a *service session* (see Chapter 4); i.e. the client service session establishes and manages the communication session to realize service level connections.

[4] This characterization applies, in general, to any kind of session.

Setup and management of stream flow connections involve setup and management of network flow connections and terminal flow connections that make up the stream flow connections. This is done in the following way. A communication session is divided into two parts: a terminal and a network. The terminal part is called a *terminal communication session* and it exists in each terminal (CPE) involved in the communication session, i.e. each CPE that contains one or more SFEPs of one or more stream flow connections that are part of the communication session. A terminal communication session is an environment (that exists in a CPE) for establishing and managing a number of terminal flow connections that are part of a communication session. The network part is called a *connectivity session* that exists (somewhere) in the domain of the *connectivity provider* that manages the connectivity session. To establish and control the network flow connections that support the stream flow connections in a communication session, the communication session needs to establish a connectivity session. To do this, the communication session determines the connectivity provider that can support the network flow connections, and requests a connectivity session. It then uses this connectivity session to establish and manage the network flow connections. Similarly, it interacts with the terminal communication sessions to set up and manage the terminal flow connections.

- *Connectivity session level*: The connection management function provided at this level is setup and management of network flow connections. Communication session level components are the clients of this level. This functionality is provided using the concept of a *connectivity session*. A connectivity session is an environment for establishing and managing a number of network flow connections. It exists in the domain of a connectivity provider. The client of a connectivity session is a *communication session*; i.e. the client communication session establishes and manages the connectivity session to set up and manage network flow connections. Setup of a network flow connection involves trails in one or more layer networks. More than one trail needs to be set up in the case where the end-points of the network flow connection are associated with different layer networks. Thus, the connection management functionality at this level consists of mapping network flow connection setup and management requests to corresponding trail management requests.
- *Layer network level*: The connection management function provided at this level is set up and management of trails and tandem connections. Connectivity session level components are the clients of this level. A trail supporting a network flow connection may span multiple-layer network domains. In such a case, connection management components in the different layer network domains interact and cooperate to establish and manage the trail. Hiding these interactions (referred to as *layer network federation*) from connectivity session level is an important aspect of the layer network level. A trail spanning multiple-layer network domains is set up in the following way. When a connectivity session determines that a trail is needed to support a network flow connection, it requests the layer network level in the

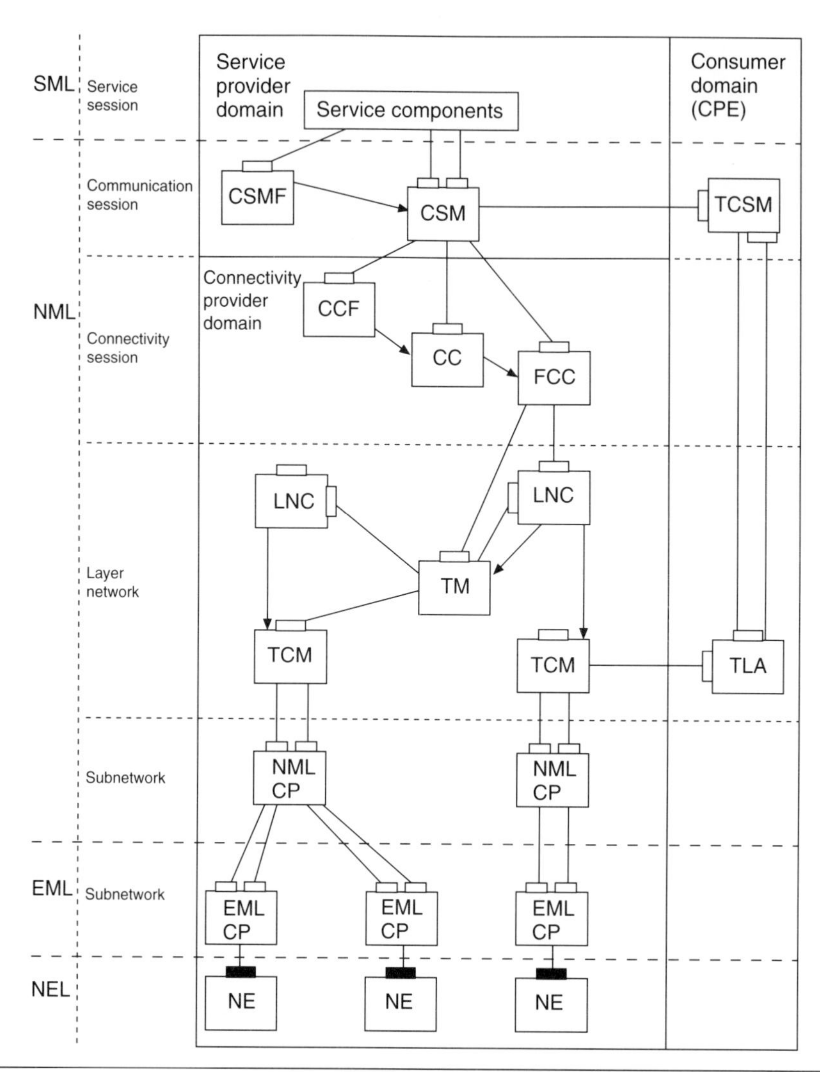

Key:

CSMF = Communication session manager factory
CSM = Communication session manager
TCSM = Terminal communication session manager
CC = Connection coordinator
CCF = Connection coordinator factory
FCC = Flow connection controller
LNC = Layer network coordinator
TM = Trail manager
TCM = Tandem connection manager
TLA = Terminal layer adaptor
CP = Connection performer
NE = Network element
SML = Service management layer
NML = Network management layer
EML = Element management layer
NEL = Network element layer
CPE = Customer premises equipment

Figure 5.13 Connection management levels and components

same connectivity provider domain (hereafter referred to as the originator domain) to set up the trail, identifying the end-points. (The end-point identification is elaborated in a later section). The layer network level in the originator domain decomposes the trail into two components: a local tandem connection in its domain and a foreign tandem connection that spans one or more foreign layer network domains. It sets up the local tandem connection using the services of the subnetwork level (see below). To set up the foreign tandem connection, it requests the neighbor layer network domain in which the tandem connection originates. The latter layer network domain, in turn, recursively decomposes the tandem connection into a local part and a foreign part and sets up the tandem connection in a similar manner.

A tandem connection that is an extreme portion of a trail has an end-point, a network trail termination point (NWTTP), that is resident in a CPE. Setting up such a tandem connection requires creation of the NWTTP. In TINA, the function of setting up and managing NWTTPs in a CPE is associated with a layer network level component residing in the CPE. Thus, setup and management of tandem connections that are the extreme portions of a trail involve interactions between the terminal components and the network components of layer network level. Hiding these interactions from the connectivity session level is also an important aspect of the layer network level.

- *Subnetwork level*: The connection management function provided at this level is set up and management of subnetwork connections. Recall that a layer network domain consists of a top-level subnetwork that may be partitioned into component subnetworks and links. This partitioning may occur in several levels until each subnetwork maps to an individual network element. Thus, setting up a subnetwork connection involves setting up subnetwork connections in the next lower level and link connections that connect these subnetwork connections. Setup and management of such subnetwork connections and link connections is the functionality associated with this level. Associated with each level of subnetwork decomposition is a subnetwork level connection management function.

5.3.2 Connection management components

Associated with each connection management level described in the previous section is a set of computational objects that provide the functionality associated with the level. These computational objects and the client–server relationships between them are also illustrated in Figure 5.13. A brief description of these objects is provided in this section. Interactions between these objects are illustrated in section 5.3.3 using an example.

5.3.2.1 *Communication session level components*

The functionality of communication session level is provided using three types of objects as described below. These objects exist in the domain of a service provider.

- *Communication session manager factory* (CSMF)

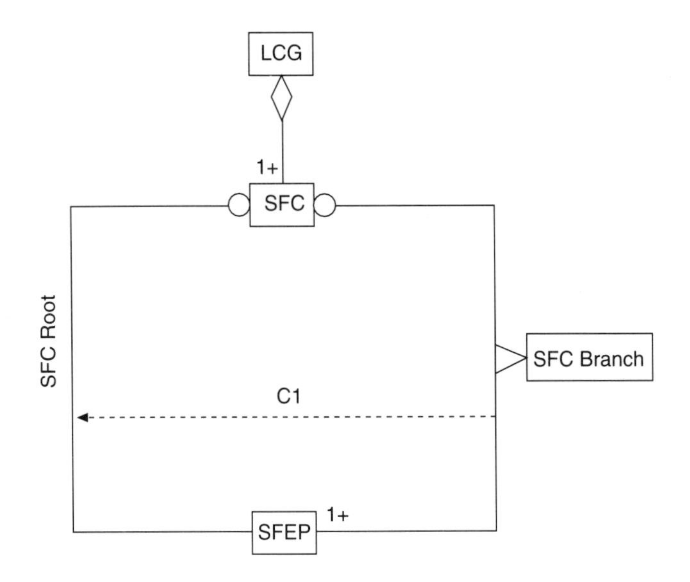

Figure 5.14 OMT diagram for a logical connection graph

- *Communication session manager* (CSM)
- *Terminal communication session manager* (TCSM)

5.3.2.1.1 Communication session manager factory

One instance of this object type exists in a service provider domain. It provides communication session establishment and release functions. At the time of a communication session setup, a client can request setup of one or more stream flow connections. For each stream flow connection, the following information is provided by the client:

- Stream flow connection name
- Root SFEP descriptor (root SFEP is the source end-point of the stream flow connection)
- Leaf SFEP descriptors
- Bandwidth and QoS parameters

The information structure that represents such information about all stream flow connections of a communication session is called a *logical connection graph*. Figure 5.14 shows the OMT diagram for a logical connection graph.

A SFEP descriptor includes information that identifies the SFEP (i.e. the application level end-point) as well as information that identifies the potential NFEPs to which the SFEP can be bound via a terminal flow connection. The latter information is provided by identifying one or more NFEP pools. A *NFEP pool* is a group of link termination points (LTPs) located in a CPE. These LTPs may be associated with different layer networks. As an example, a CPE (called CPE1) is connected to a frame relay switch and two ATM switches. The three LTPs, one corresponding to the link to each switch, can be grouped to form a NFEP pool. When a service session requests a communication session setup, it does not prescribe the NFEPs to which each SFEP must be bound but identifies only the NFEP pools. The selection of the NFEP for each terminal flow connection is made later in the layer network level interactions between the CPE and the network when a tandem connection is set up. This "late binding" of SFEP to NFEP allows a flexible selection of NFEPs that takes into account the connectivity layer network topology and QoS/bandwidth availability information on the various layer networks. For example, in the scenario referred to earlier, the NFEP selection should consider the following aspects:

- If the NFEP chosen is associated with the ATM network, is the other CPE also attached to the ATM network? If not, are there interworking units providing adaptation of ATM layer information to the characteristic information of the layer network to which the other CPE is attached? This is an issue concerning connectivity layer network topology and thus a concern in connectivity session level. This should be transparent at communication session level and to service level applications.
- Suppose both CPEs are attached to the ATM network. Now, the selection of NFEP on CPE1 should take into account the bandwidth and QoS available in the ATM network. For example, it is possible the required bandwidth and QoS can be met only if the NFEP chosen is in the LTP attached to ATM switch A, but not in the LTP attached to switch B. This is a traffic and bandwidth management concern in the layer network level. This should be transparent to higher levels.

Late binding of SFEPs to NFEPs honours these separation concerns and enables a flexible selection of NFEPs.

When a communication session setup request is received, the CSMF either creates a new CSM object for controlling the new communication session or assigns the control responsibility to an existing CSM. This decision is based on the scheduling policy used by the CSMF. In either case, the CSMF delegates to the CSM object the communication setup request and passes along the communication session information received from the client.

5.3.2.1.2 Communication session manager

One or more instance of this object type exists in a service provider domain. Each CSM object manages one or more communication sessions.[5] For each communication session under its control, the CSM offers the following management capabilities:

[5] If the CSM manages multiple communication sessions, the management operations for each session are offered via a separate computational interface.

- Stream flow connections setup
- Addition/removal of branches to/from point-to-multipoint stream flow connections. A branch of a SFC represents the point-to-point connectivity from the root SFEP to a leaf SFEP.
- Modification of bandwidth and QoS parameters of stream flow connections
- Activation of stream flow connections or their branches: the activation operation on a SFC sets the *administrative state* of the SFC to an *unlocked* state. A SFC transports information only if its administrative state is unlocked. In the case of a point-to-multipoint SFC, activation can be applied to specific branches.
- Deactivation of stream flow connections or their branches: the deactivation operation on a SFC sets the administrative state of the SFC to a *locked* state. A SFC does not transport information if its administrative state is locked. In the case of a point-to-multipoint SFC, activation can be applied to specific branches.
- Stream flow connections release
- SFC failure/recovery notifications: when a SFC fails or recovers from a failure, the CSM communicates this event to the client service session.

When a CSM is requested to set up a SFC (either as part of a communication session setup or by an explicit request), the CSM interacts with the TCSM resident in each CPE that is a participant to the SFC. In this interaction, the CSM sends to each TCSM a *correlation identifier*. The purpose of this is as follows. Recall that at SFC setup time NFEPs are not chosen. NFEPs are selected later when the boundary tandem connections are set -up in the layer network level. Only after the NFEPs have been chosen can the TCSMs set up the TFCs associated with the SFC. When a NFEP is chosen in a CPE, the layer network level component in the CPE, called the *terminal layer adapter* (TLA) (see below) notifies the TCSM that the NFEP has been set up passing along the correlation identifier. Upon receipt of this notification, the TCSM sets up the TFC between the SFEP and chosen NFEP.

When a CSM is requested to set up a SFC, it sends to each TCSM the correlation identifier as described above. After this step, the CSM interacts with a connectivity session level component to set up the connectivity session for the communication session.

5.3.2.1.3 Terminal communication session manager

One instance of this object type exists in each CPE. The TCSM object in a CPE manages all terminal communication sessions that exist in the CPE. The TCSM offers the following management operations:

- SFEP registration: This operation is invoked by a user application when it decides to participate in a stream binding with other applications as part of a service session. The application invokes this function to make the TCSM aware of the SFEP that the application wishes to use in the stream binding. Note that SFEP creation is application responsibility. Upon receipt of this invocation, the

TCSM determines the NFEP pools that can be associated with the SFEP, includes this information in a SFEP descriptor and returns it to the application as a result.

- Set correlation identifier: This operation is invoked by a CSM to send a correlation identifier to the TCSM that is associated with a TFC. The TFC will be set up by the TCSM later when the TLA informs the TCSM of the NFEP selection for the TFC. The correlation identifier contains the names of the SFC and the NFC that are associated with the TFC.
- NFEP association with TFC: This operation is invoked by the terminal layer adapter (TLA) (see below) after it chooses a NFEP for a TFC (as part of boundary tandem connection setup). The TLA passes to TCSM the NFEP identification and the correlation identifier. Upon receipt of this invocation, the TCSM uses the correlation identifier to determine the SFEP to be bound to the TFC, establishes the TFC between the SFEP and the chosen NFEP, and sends a confirmation response to the TLA. Establishment of a TFC involves details specific to the network technology of the chosen NFEP and characteristics of the device associated with the SFEP, and may include setting up adaptation and buffering functions.
- NFEP disassociation from TFC: This operation is invoked by a CSM to release a TFC. The CSM issues this request when it either releases a SFC or changes the SFC–NFC mapping.
- SFEP deregistration: This operation is invoked by the user application when a SFEP it created ceases to be a participant in a stream binding.
- Remove correlation identifier: This operation is invoked by a CSM to release a TFC. The CSM issues this request when it releases a SFC.
- TFC activation/deactivation: This operation is invoked by a CSM to activate/deactivate a TFC. The CSM issues this request when it activates/deactivates a SFC.
- Modification of TFC bandwidth and QoS parameters: This operation is invoked by a CSM to modify a TFC.
- TFC failure/recovery notifications: When a TFC fails or recovers from a failure, the TCSM communicates this event to the associated CSM.

5.3.2.2 Connectivity session level components

The functionality of connectivity session level is provided using three types of objects as described below. These objects exist in the domain of a connectivity provider.

- *Connection co-ordinator factory* (CCF)
- *Connection co-ordinator* (CC)
- *Flow connection controller* (FCC)

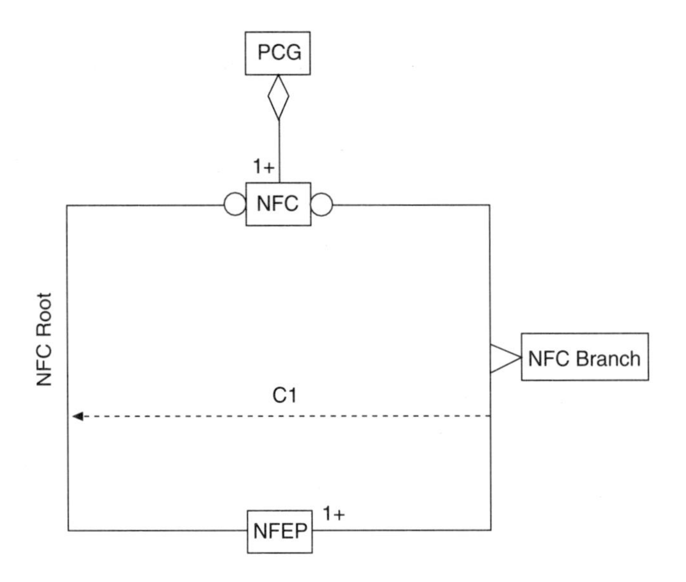

Figure 5.15 OMT diagram for a physical connection graph

5.3.2.2.1 Connection co-ordinator factory

One instance of this type exists in a connectivity provider domain. It provides connectivity session establishment and release functions. At the time of a connectivity session setup, a client can request setup of one or more network flow connections. For each network flow connection, the following information is provided by the client:

- Network flow connection name
- Root NFEP descriptor
- Leaf NFEP descriptors
- Correlation identifier

The information structure that represents such information about all network flow connections of a connectivity session is called a *physical connection graph*. Figure 5.15 shows the OMT diagram for this graph. A NFEP descriptor includes NFEP pool identification and the bandwidth and QoS parameters associated with the NFEP.

Upon receipt of a connectivity session setup request, the CCF either creates a new CC object for controlling the new connectivity session or assigns the control responsibility to an existing CC, depending on the scheduling policy used by the CCF. In either case, the

CCF delegates to the CC object the communication setup request and passes along the connectivity session information received from the client.

5.3.2.2.2 Connection co-ordinator

One or more instance of this object type exists in a connectivity provider domain. Each CC object manages one or more connectivity sessions. For each connectivity session under its control the CC offers the following management capabilities:

- Network flow connections setup
- Addition/removal of branches to/from point-to-multipoint network flow connections.
- Modification of bandwidth and QoS parameters of network flow connections
- Activation of network flow connections or their branches: the activation operation on a NFC sets the *administrative state* of the NFC to an *unlocked* state. A NFC transports information only if its administrative state is unlocked. In the case of a point-to-multipoint NFC, activation can be applied to specific branches.
- Deactivation of network flow connections or their branches: the deactivation operation on a NFC sets the administrative state of the NFC to a *locked* state. A NFC does not transport information if its administrative state is locked. In the case of a point-to-multipoint NFC, activation can be applied to specific branches.
- Network flow connections release

When a CC is requested to set up a NFC, it creates a flow connection controller (FCC) object and delegates the setup request to that object. Thus, associated with each NFC is a FCC object that provides management operations for the NFC. Similarly, when the CC receives a request for a management operation on a set of NFCs, it passes the request to the FCC associated with each NFC in the set. The scheme of using a separate FCC object for controlling each NFC is used to allow maximum degree of distribution. These objects can be grouped into one cluster at the engineering level for performance reasons, if necessary.

5.3.2.2.3 Flow connection controller

One instance of this object type is associated with each NFC. A FCC object provides the following management operations for the NFC under its control:

- Network flow connection setup
- Addition/removal of branches if the NFC is point-to-multipoint connection
- Modification of bandwidth and QoS parameters of the network flow connection
- Activation of the network flow connection or its branches
- Deactivation of the network flow connection or its branches
- Network flow connection release

When a FCC is requested to set up a NFC it determines how many trails are to be set up to provision the NFC. This determination is based on the layer networks to which the NFEP pools of the root and leaf NFEPs belong and the topology of the connectivity layer network. In general, a sequence of trails may have to be set up, with boundary trails terminating in the CPEs participating in the NFC and intermediate trails originating and terminating in the interior of the connectivity layer network. To simplify discussion, here we assume that only one trail needs to be set up to provision a NFC; i.e. no peer-to-peer layer interworking is needed. To set up the trail, the FCC requests the layer network co-ordinator that controls the layer network to which the NFEP pools belong. Before it issues the trail setup request, the FCC performs the following mappings:

- Each NFEP pool is mapped to a LTP on the chosen layer network. Recall that an LTP is an aggregate of TLTPs (i.e. UNI link terminations).
- The bandwidth and QoS parameters of the NFEPs are mapped to technology-specific bandwidth and QoS parameters.

Similarly, other NFC management operations are mapped into trail management operations taking into account layer network specific details.

5.3.2.3 Layer network level components

The functionality of layer network level is provided using four types of objects as described below. These objects exist in the domain of a connectivity provider.

- *Layer network co-ordinator* (LNC)
- *Trail manager* (TM)
- *Tandem connection manager* (TCM)
- *Terminal layer adapter* (TLA)

5.3.2.3.1 Layer network co-ordinator

One instance of this object type exists in a connectivity provider domain for each layer network domain (LND) contained in the connectivity provider domain. It provides trail setup and release operations to the FCC objects in the same connectivity provider domain. Further, it provides tandem connection setup and release operations to the LNC of each neighbor foreign LND, and also requests them to perform tandem connection setup and release operations. Thus, an LNC has a peer-to-peer relationship (both client and server roles) with each neighbor LNC.

When it receives a trail setup request, the LNC creates a TM object for the management of the new trail and passes the trail setup request to the TM. Similarly, when it receives a tandem connection setup request, the LNC creates a TCM object for the management of the new tandem and passes the tandem connection management request to the TCM object.

5.3.2.3.2 Trail manager

One instance of this object type exists for each trail in the originating connectivity provider domain, i.e. the domain in which the FCC that requested the trail setup exists. A TM object provides the following management operations for the trail under its control:

- Trail setup: this operation is invoked by the LNC.
- Addition/removal of branches if the trail is point-to-multipoint connection: this operation is invoked by the FCC.
- Modification of bandwidth and QoS parameters of the trail: this operation is invoked by the FCC.
- Activation of the trail or its branches: this operation is invoked by the FCC.
- Deactivation of the trail or its branches: this operation is invoked by the FCC.
- Trail release: this operation is invoked by the LNC.

When a TM is requested to set up a trail, it determines the layer network domains that are to be traversed by the trail. This determination is based on the layer network domains to which the LTPs of the root and leaf NFEPs belong, the topology of the layer network, and the route selection policy of the TM.[6] Based on this determination, the TM decomposes the trail into two components: a tandem connection in the local LND and zero or more foreign tandem connections. (More than one foreign tandem connection may need to be set up in the case of a multipoint trail.) Each foreign tandem connection has one end-point in a neighbor LND and the other end-points are in the LTPs associated with one or more leaf end-points of the trail. To set up the tandem connection in the local LND, the TM requests the LNC of the local LND.[7] To set up a foreign tandem connection, the TM requests the LNC of the neighbor foreign LND associated with the tandem connection.

Similarly, the TM maps other trail management operations into one or more tandem connection management operations depending on the trail branches affected by the operation, and requests the corresponding TCMs to perform the management operation.

5.3.2.3.3 Tandem connection manager

One instance of this object type exists for each tandem connection originating in the connectivity provider domain. A TCM object provides the following management operations for the tandem connection under its control:

- Tandem connection setup: this operation is invoked by the LNC.
- Addition/removal of branches if the tandem connection is point-to-multipoint connection: this operation is invoked either by the TM in the local domain or by a TCM in a foreign connectivity provider domain.

[6] Source routing is a possibility.

[7] Notice that when a LNC requests a TM to set up a trail, the TM calls back the LNC to set up the local tandem connection. To avoid deadlocks, either the LNC should make non-blocking invocations or the LNC implementation should be multithreaded.

- Modification of bandwidth and QoS parameters of the tandem connection: this operation is invoked either by the TM in the local domain or by a TCM in a foreign connectivity provider domain.
- Activation of the tandem connection or its branches: this operation is invoked either by the TM in the local domain or by a TCM in a foreign connectivity provider domain.
- Deactivation of the tandem connection or its branches: this operation is invoked either by the TM in the local domain or by a TCM in a foreign connectivity provider domain.
- Tandem connection release: this operation is invoked either by the TM in the local domain or by a TCM in a foreign connectivity provider domain.

When a TCM is requested to set up a tandem connection, it determines the layer network domains that are to be traversed by the tandem connection. This determination is based on the layer network domains to which the LTPs associated with the end-points of the tandem connection belong, the topology of the layer network, and the route selection policy of the TCM (if source routing was not used for the trail). Based on this determination, the TCM decomposes the tandem connection into two components: a tandem connection in the local LND and zero or more foreign tandem connections. (More than one foreign tandem connection may need to be set up in the case of a multipoint trail.) To set up a foreign tandem connection, the TCM requests the LNC of the neighbor foreign LND associated with the tandem connection. Before it sets up the tandem connection in the local LND, the TCM determines if the local tandem connection is a boundary tandem connection, i.e. one of the end-points is on a CPE. If so, it requests the terminal layer adapter object associated with the CPE to set up the NWCTP and NWTTP for the trail end-point on the CPE. This request causes the TLA to assign bandwidth and channel numbers (such as VPI/VCI in the case of ATM) on the associated access link. Upon receiving a success response from the TLA, the TCM proceeds to set up the local tandem connection.

To set up the local tandem connection, the TCM selects the route in terms of (second level) subnetworks and links in the local LND. It assigns bandwidth and channel number on each link in the chosen route and requests the network management layer connection performer (NML-CP) associated with each subnetwork in the chosen route to set up the associated subnetwork connection.

Similarly, the TCM maps other tandem connection management operations into operations on foreign tandem connections, if necessary, and subnetwork connections in the local LND. Then, it requests the TCMs in the foreign domains and/or NML-CPs in the local domain to perform the corresponding management operation.

5.3.2.4 Terminal layer adapter

One instance of this object type exists in a CPE for the layer network to which the CPE is attached. This object serves two purposes:

- It manages the trail terminations occurring on the CPE. The management includes creation, modification, and release. It performs these operations upon requests from a TCM in the associated LND. These operations involve bandwidth management and channel assignment of access links that connect the CPE to adjoining NEs.

- It serves as the linkage point between the technology-independent and technology-specific aspects of a stream flow connection setup. After it sets up a NWTTP for a trail associated with a stream flow connection, the TLA notifies the TCSM that a NFEP (which is a generic view of the NWTTP) has been set up. Upon receiving this notification, the TCSM sets up the terminal flow connection that binds the SFEP to the chosen NFEP.

5.3.2.5 Subnetwork level components

The functionality of subnetwork level is provided using two types of objects as described below. These objects exist in the domain of a connectivity provider.

- Network management layer connection performer (NML-CP)
- Element management layer connection performer (EML-CP)

5.3.2.5.1 Network management layer connection performer
One instance of this object type exists in a connectivity provider domain for each non-element subnetwork in the domain. This object provides operations for setup, modification, and release of subnetwork connections in the associated subnetwork. When a subnetwork connection is requested, this object determines the route for the subnetwork connection, in terms of lower-level subnetworks and links that interconnect these subnetworks. It assigns bandwidth and channel number on each link in the chosen route and requests either the NML-CP or the EML-CP associated with each subnetwork in the chosen route to set up the associated subnetwork connection.

Similarly, the NM-CP maps other subnetwork connection management operations into operations on lower-level subnetwork connections. Then it requests either the NML-CP or the EML-CP of the associated subnetwork to perform the corresponding management operation.

5.3.2.5.2 Element management layer connection performer
One instance of this object type exists in a connectivity provider domain for each element subnetwork in the domain. An element subnetwork is one that is not further decomposed into lower-level subnetworks and links, and thus represents an individual NE. This object provides operations for setup, modification, and release of subnetwork connections in the associated NE. When a subnetwork connection management operation is requested, this object performs the operation by interacting with the NE controller using an interface that may be NE-specific.

5.3.3 A communication session setup scenario

This section describes an illustrative example of a communication session setup. The scenario described here is related to the service example described in section 4.7 in the following manner. Section 4.7.2 described how a multimedia conferencing service between a user (Franz) and a travel agent (saleswoman) is set up. From the NRA perspective, setting up this service involves setting up a communication session with the following stream flow connections:

- A point-to-point SFC, SFC1, with the source SFEP corresponding to the audio source on Franz's workstation and the sink SFEP corresponding to the audio sink on the bridge.
- A point-to-point SFC, SFC2, with the source SFEP corresponding to the video source on Franz's workstation and the sink SFEP to the video sink on the bridge.
- A point-to-point SFC, SFC3, with the source SFEP corresponding to the audio source on the travel agent's workstation and the sink SFEP to the audio sink on the bridge.
- A point-to-point SFC, SFC4, with the source SFEP corresponding to the video source on the travel agent's workstation and the sink SFEP to the video sink on the bridge.
- A point-to-multipoint SFC, SFC5, with the source SFEP corresponding to the audio source on the bridge and two sink SFEPs, one corresponding to the audio sink on Franz's workstation and another to the audio sink on the travel agent's workstation.
- A point-to-multipoint SFC, SFC6, with the source SFEP corresponding to the video source on the bridge and two sink SFEPs, one corresponding to the video sink on Franz's workstation, and another to the video sink on the travel agent's workstation.

This section describes how the above communication session is set up outlining the various steps involved and the interactions between the different components of the connection management architecture. To simplify the description, the following simple connectivity layer topology is assumed. The connectivity layer network consists of only one layer network and the entire layer network is under the control of one connectivity provider; i.e. there is only one LND. The LND has a three-tier subnetwork decomposition; the top-level subnetwork is composed of several second-level subnetworks interconnected by links; each second-level subnetwork is composed of a third-level subnetwork interconnected by links; each third-level subnetwork maps to an NE.

1. The service session manager (SSM) of the multimedia conferencing service requests the CSMF to set up a communication session with the six SFCs as described above. For each SFC to be created, the SSM specifies each SFEP using a SFEP descriptor that identifies the NFEP pool (set of LTPs) from which the NFEP can be selected by the connection management function.
2. CSMF instantiates a CSM object to manage the communication session and passes the communication session setup request to the CSM.

3. For each SFC to be created, the CSM creates a correlation identifier. It sends the correlation identifiers for SFC1, SFC2, SFC5, and SFC6 to the TCSM residing in Franz's workstation. Similarly, it passes the correlation identifiers for SFC3, SFC4, SFC5, and SFC6 to the TCSM residing in the travel agent's workstation.
4. The CSM requests the CCF in the connectivity provider domain (with whom the retailer already has a business relationship) to set up a connectivity session with six NFCs, one for each SFC. For each NFC to be created, the CSM specifies each NFEP using a NFEP descriptor that identifies the NFEP pool (set of LTPs) from which the NFEP can be selected by the connectivity provider.
5. CCF instantiates a CC object to manage the connectivity session and passes the connectivity session setup request to the CC.
6. The CC instantiates a FCC object to manage each network flow connection and passes the NFC information to the FCCs.
7. Each FCC determines how many trails are to be set up for the NFC. In this example, it is assumed that all communicating parties are attached to the same layer network and thus only one trail is needed for each NFC.
8. Each FCC requests the LNC of the local LND to set up the trail. The FCC identifies for each trail end-point (NWTTP) the set of LTPs from which the NWTTP can be selected.
9. The LNC handles each trail set up independently. For brevity, the setup of only one trail is described here. The LNC creates a TM object to manage each trail and passes the trail setup request to the TM.
10. The TM selects the LTP for each end-point of the trail (based on the traffic information known to it) and computes the route for the trail in terms of LNDs to be traversed by the trail. In this example, it is assumed that all end-points are in the local LND. TM requests the LNC to set up a tandem connection in the local LND identifying the LTP for each end-point.
11. The LNC creates a TCM object and passes the tandem connection setup information to the TCM.
12. The TCM determines that all end-points of the tandem connection are in the CPE. For each end-point, it requests the TLA on the associated CPE to create a NWTTP on the selected LTP.
13. Each TLA creates a NWTTP and notifies to the TCSM in the CPE the creation of the NWTTP (and thus the NFEP) for a SFC passing along the correlation identifier. (This requires that correlation identifiers be passed in all interactions above the layer network level.)
14. Each TCSM sets up the TFC for the SFC identified by the correlation identifier and sends a success response to the TLA.
15. Each TLA in turn sends a success response to the TCM. When the TCM

receives success responses from all TLAs, it proceeds to build the tandem connection in the local LND.

16. The TCM computes the route for the tandem connection in terms of second-level subnetworks and links in the local LND that are to be traversed by the tandem connection. On each end-point of a link in the chosen route for the tandem connection, the TCM creates a NWCTP by allocating bandwidth and channel number and creates an associated edge object. The TCM requests the NML-CP of each second-level subnetwork to be traversed by the tandem connection to create a subnetwork connection (SNC) identifying the edges of the SNC.
17. Each NML-CP computes the route for the subnetwork connection in terms of the third-level subnetworks and links in the local LND that are to be traversed by the subnetwork connection. On each end-point of a link in the chosen route for the subnetwork connection the NML-CP creates a NWCTP by allocating bandwidth and channel number and creates an associated edge object. The NML-CP requests the EML-CP of each third-level subnetwork to be traversed by the tandem connection to create a subnetwork connection (SNC) identifying the edges of the SNC.
18. Each EML-CP sets up the cross-connection in the NE and sends a success response to the NML-CP that requested the connection.
19. When a NML-CP receives success responses from all EML-CPs, it sends a success response to the TCM that requested the subnetwork connection.

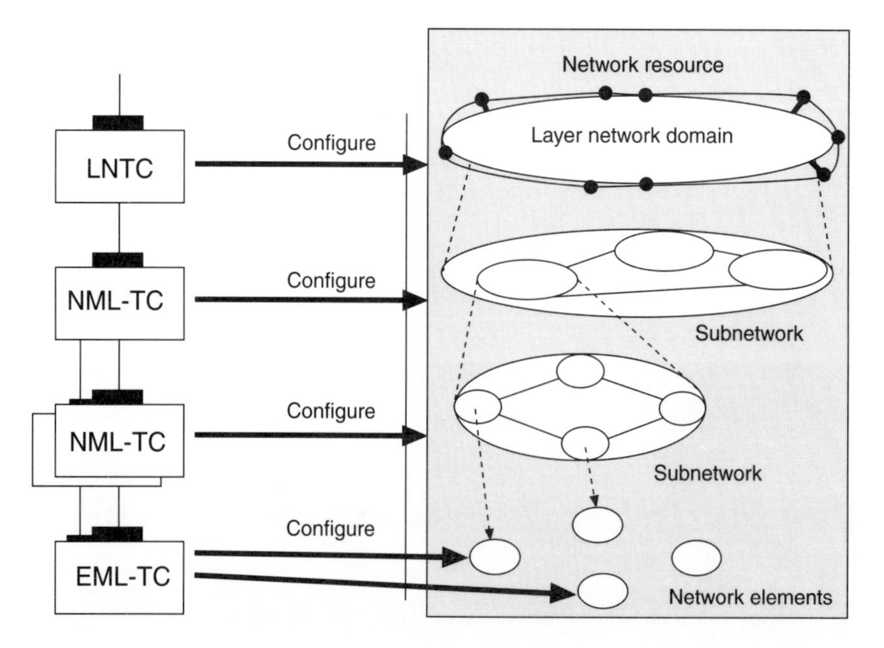

Figure 5.16 Network topology configuration management computational architecture

20. When a TCM receives success responses from all NML-CPs, it sends a success response to the LNC.
21. The LNC sends a success response to the TM that requested the tandem connection.
22. The TM sends a success response to the LNC that requested the trail.
23. The LNC sends a success response to the FCC that requested the trail.
24. The FCC sends a success response to the CC.
25. When the CC receives success responses from all FCCs, it sends a success response to the CCF.
26. The CCF sends a success response to the CSM.
27. The CSM sends a success response to the CSMF.
28. The CSMF sends a success response to the SSM.
29. The communication session setup is complete.

5.4 TOPOLOGY CONFIGURATION MANAGEMENT ARCHITECTURE

The goal of network topology configuration management (NTCM) in NRA is to manage the lifecycle of network topology resources (such as subnetworks, links, layer network domains). The NTCM functions can be classified as follows:

- *Installation support*: this consists of setting up the initial configuration of a topological resource. Installation of NEs (physical equipment) is technology- and equipment-specific and is thus outside the scope of NTCM. After an NE has been installed, configuring the NE as an element subnetwork (including LTP/port configuration) either through self-discovery or human user interaction is within the scope of NTCM. An interesting aspect of NTCM is that the initial configuration of a topology resource (such as a subnetwork) includes also setting up the necessary management software (such as NML-CP to manage connections in the subnetwork, fault-management components to manage faults in the subnetwork, etc.). Thus, installation support initializes the topology resources and makes them available for use by other management functions.
- *Provisioning*: this consists of managing changes in the configuration of a topology resource, such as addition or removal of links between component subnetworks in a subnetwork, changes in link capacities and addition or removal of ports in element subnetworks. When configuration changes are made to a topology resource, the NTCM component responsible for the resource notifies these changes to other management components so that they can start/stop the resource usage as the case may be.
- *Status and control*: this consists of monitoring the operational state and controlling the administrative state of topology resources. When the operational state of a topology resource changes, e.g. from operational to failed, the NTCM records

the change and notifies other management components. The NTCM is responsible for managing changes to the administrative state of a topology resource, e.g. the change from unlocked to locked, for testing, fault diagnosis or some other maintenance purpose. When such administrative state changes occur, the NTCM notifies other management components. Thus, NTCM serves as the steward of the topology view of the network, and other management components depend on this topology view for performing their functions.

The above-described scope of NTCM functions is very much the same as that ascribed to configuration management functions in the network management literature. The significant difference is that the scope of NTCM is restricted to network topology management and NE installation management is outside the scope of NTCM.

Figure 5.16 illustrates the computational architecture of NTCM. As in the case of connection management, the NTCM components are also categorized into several levels as follows:

- *Layer network topology configurator* (LNTC): One such object exists for each layer network domain and it manages the topology of the associated LND. This involves managing the topology of the LND in terms of the second-level subnetworks that make up the top-level subnetwork of the LND, provisioning of links in the LND that terminate on CPEs, provisioning of links that connect the LND with foreign LNDs of the same layer network and provisioning of layer interworking units that connect the LND with a local or foreign LND of another layer network. The LNTC is responsible for creating and controlling the computational objects that deal with other FCAPS management functions for the LND (such as LNC).
- *Network management layer topology configurator* (NML-TC): One such object exists for each non-element subnetwork in the local LND. This object manages the topology of the associated subnetwork. This involves managing the topology of the subnetwork in terms of its component subnetworks and links. The NML-TC is responsible for creating and controlling the computational objects that deal with other FCAPS management functions for the subnetwork (such as NML-CP).
- *Element management layer topology configurator* (EML-TC): One such object exists for each element subnetwork in the local LND. This object manages the configuration of the associated element subnetwork (NE). The EML-TC is responsible for creating and controlling the computational objects that deal with other FCAPS management functions for the element subnetwork (such as EML-CP).

5.5 FAULT-MANAGEMENT ARCHITECTURE

Network resource fault management is concerned with detection, localization, and correc-

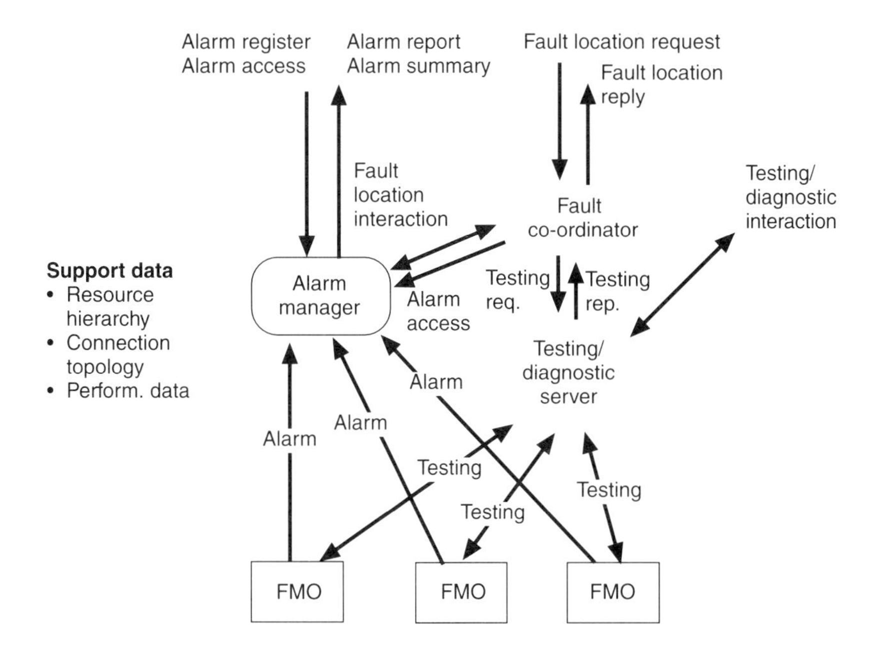

Figure 5.17 Fault-management computational architecture framework

tion of faults (i.e. abnormal behavior) of network resources. In the network management literature, the following activities are associated with fault management [AIDA94]:

- *Alarm surveillance*: this activity receives alarm reports from network resources, filters alarms based on filtering conditions that include alarm severity, records filtered alarms, initiates alarm analysis to detect the root cause of the filtered alarms, and reports the root cause to higher-level management functions.
- *Fault localization*: this activity initiates testing and diagnostic procedures to identify the resource component or module that is faulty.
- *Testing*: this activity is performed in support of fault localization. It ranges from performing simple tests that verify the functionality of a resource to complex analysis of a set of test results to determine the faulty component.
- *Fault correction*: this activity is responsible for the repair of the faulty component (including dispatch of workforce to handle the repair) and initiation of procedures for activating redundant resources, if any, to replace the failed component.
- *Trouble administration*: this is a support activity that enables human users and management functions to keep track of problems reported and the progress of the resolution activities.

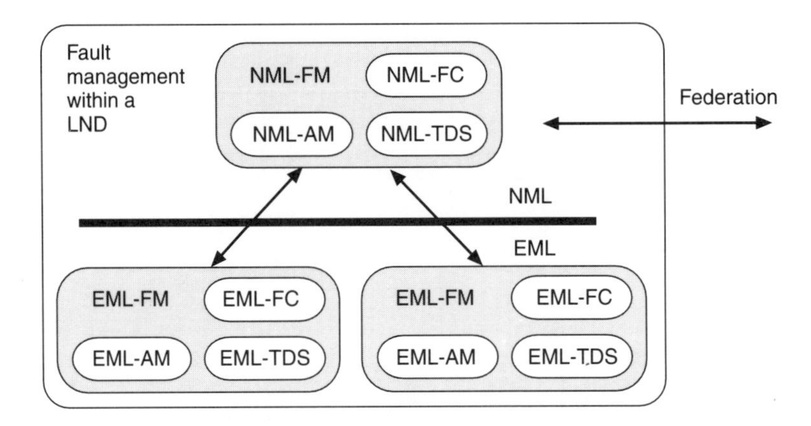

Figure 5.18 Fault-management computational architecture

Among the above fault management activities, the current version of NRA has addressed only the first three and only these are described below.

Figure 5.17 illustrates the computational architecture framework used for fault management in NRA. Figure 5-18 illustrates the use of this framework for fault management in the NML and EML layers.

The fault-management computational architecture framework consists of three computational objects as described below (see Figure 5.18):

- *Alarm manager* (AM): This object receives alarm reports from resources under its purview (NE level resources in the case of EML and subnetwork level resources in the case of NML). It records these reports in an event log and applies filtering conditions previously set by higher-level management functions or human administrators. It passes the filtered alarms to the *fault co-ordinator* (FC) object to initiate root cause analysis of the alarms. If it receives any fault location results from FC it reports the fault to higher-level management functions and administrators that have subscribed to such reports. Otherwise, it reports the uncorrelated alarms to higher-level management functions and administrators that have subscribed to such reports. Note that a FC may not be able to locate faults in resources that are beyond its span of control. For example, a NML fault-management function dealing with a subnetwork cannot locate faults in a topological link that connects the subnetwork to another subnetwork. Such faults can be located in the higher-level NML.
- *Fault co-ordinator* (FC): This object receives alarms from the AM and performs the root cause analysis on the received alarms and determines, if possible, the resource component that is faulty. To perform this analysis, the FC

uses different kinds of support data, such as network topology (i.e. resource hierarchy) data, data on the different connectivity resources, and current alarm records. Further, it initiates tests and diagnostics on related resources and analyzes the test results to locate the fault. If the fault is located, it reports this to the AM.

- *Testing and diagnostic server* (TDS): This object receives requests for tests and diagnostics to be performed on one or more resources under its purview, performs these procedures, and reports the results to the requester. Testing and diagnostic requests may be received from either the FC object in the same management domain, fault-management objects in a higher-level management domain, or fault-management objects in a foreign layer network domain. The last aspect mentioned is related to the issue of layer network federation for fault management and this problem remains to be addressed in NRA.

5.6 ACCOUNTING MANAGEMENT ARCHITECTURE

Broadly, the problem of accounting management in telecommunications networks has two aspects:

- *Usage measurement*: Collection of resource usage statistics on a service basis; i.e. duration and amount of resources used in support of a service.
- *Billing services*: Computation of charges ("bills") to be paid by each subscriber based on the usage statistics of the services related to the subscriber and the billing policies for the services; delivery of billing notices and tracking payment by subscribers.

Of the above two, only the former is under the scope of accounting management in NRA. The latter is a service (or a collection of services) that may be designed much like any other TINA service. Hence, in this section discussion is restricted only to the former.

In NRA, the usage measurement aspect of accounting management is further subdivided into the following activities:

- *Usage measurement policy management*: this activity deals with setup and modification of usage measurement policies. Examples of policies are determination of resources whose usage is to be monitored (typically different types of connectivity resources such as subnetwork connections, trails, network flow connections, etc.), determination of accounting metrics for resources (i.e. what accounting information to collect) where some of the metrics may be technology dependent, and determining when to start and stop usage measurement. An example of a technology-independent accounting metric is number of Mbits transported. Examples of ATM technology-dependent metrics are ingress total cells and ingress high-priority cells.

- *Accounting data collection and management*: this activity deals with collection of accounting data, logging the collected data, and management of this data including query support and data-retention mechanisms.

Following the general NRA principle of decomposing management components into several levels reflecting the network structure, the accounting management components are also divided into several levels, such as EML, NML, layer network level, and connectivity layer level. The accounting management components in each level constitute an *accounting management domain*.

Figure 5.19 illustrates the computational architecture of an accounting management domain. Within an accounting domain, there are three types of computational objects:

- *Accounting policy manager*: One such object exists in each accounting management domain. This object maintains the accounting policy associated with the domain. The policy is represented as a set of tag value pairs, where each tag denotes a policy parameter (or rule). Examples of such parameters are: set of resource types subject to usage measurement, measurement interval, and retention duration of accounting records. The object provides operations for querying and modifying these parameters.
- *Accountable object*: An accountable object represents a resource that is subject to accounting. A computational object representing a network resource is an accountable object if it supports an interface that provides accounting control operations (start, stop, resume, set accounting interval, set accounting metrics, etc.). When an object representing a resource is created, it retrieves the accounting policy applicable to the resource from the accounting policy manager. If the accounting parameters of a domain change, the accounting policy manager communicates the new parameter values to all accountable objects in the domain through the accounting control interface of the individual object. Each accountable object sends an accounting event report on the expiration of each accounting interval specified for the resource. This report contains the usage statistics for the accounting metrics associated with the resource.
- *Metering manager*: One such object exists in each accounting management domain. This object receives accounting event reports from all accountable objects in the domain, and logs them. Further, it serves as the disseminator of event reports to other accounting management domains. An object in one domain can subscribe to event reports from another domain by requesting the metering manager of the latter domain.

5.7 NRA AND INTERNET MANAGEMENT

While the current version of NRA provides a solid foundation for network resource management in TINA, the current NRA handles only connection-oriented networks.

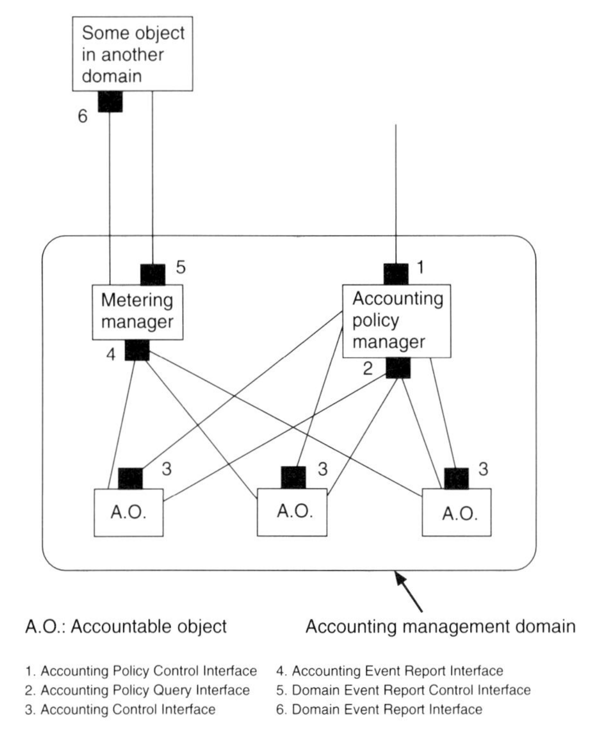

Figure 5.19 Computational architecture of an accounting management domain

Connectionless networks, of which the Internet is a prime example, have not been addressed. Given the wide and growing prevalence of the Internet, this is a significant limitation. There are two aspects to this problem: one is using the Internet as a transport network for carrying TINA streams and the other is management of Internet network resources. The TINA-C core team has developed a preliminary solution for the first problem, and this is described below.

The approach that is being explored for the problem of using the Internet to establish and manage stream flow connections is modeling the internet as a layer network. The concept of a layer network, which has been invariably used in the telecommunications networking literature to represent only connection-oriented networks, is generalized to represent also connectionless networks. Since the concept of trail is closely associated with connection-oriented networks, a new concept called *layer network binding* (LNB) is proposed that is applicable for both connection-oriented and connectionless networks. A layer network binding is an association between two or more end-points (*layer network access points*) in a layer network. In the case of the Internet, a layer network access point represents either a UDP port or a TCP port depending on the Internet protocol used for the stream flow. The concept of trail then becomes a specialization of the layer network binding concept and is used only for connection-oriented networks. The Internet is modeled as a layer network that supports different types of bindings, such as UDP binding, TCP binding, and multicast binding. These bindings vary in aspects such as topologies supported and QoS guarantees.

To support TINA streams using the Internet, the current TINA connection management architecture is modified as follows. The layer network co-ordinator (LNC) interface that is used by the flow connection controller (FCC) is generalized to support creation of LNBs. The FCC identifies the end-points of a LNB using NFEP pools. In the case of the Internet, a NFEP pool is a set of IP addresses. When it receives a LNB setup request, the LNC that manages the Internet creates a *layer network binding manager* (LNBM) to set up and manage an Internet layer network binding. A LNBM is the counterpart of the trail manager (TM) that is used for managing a trail in a connection-oriented network. To provide transparency to FCC, both LNBM and TM support an identical interface.

The other connection management component that is modified to support the Internet is the terminal layer adapter (TLA). To handle Internet transport, a TLA exists in each CPE. The TLA encapsulates the Internet protocol details and offers interfaces for LNB management, one interface for each Internet transport protocol supported in the CPE (UDP, TCP, etc.). To set up a LNB using a specific Internet transport protocol, the LNBM uses the appropriate interface offered by the TLAs in the CPEs.

For example, the TLA interface for TCP transport provides the following operations:

- *Create TCP end-point*: this operation takes as input a set of IP addresses corresponding to a subset of the network interfaces of the CPE, selects one IP address among them, creates a TCP port, and returns a fully resolved TCP address (IP address + TCP port) for the created end-point.
- *Accept*: this operation takes a set of IP addresses and optionally a remote TCP address as input. It selects one IP address, creates a TCP port, returns a fully resolved TCP address for the created end-point, and sets the CPE to accept TCP connection requests to the created local TCP port. If the remote TCP address is specified, connection requests only from the specified remote address will be accepted.
- *Connect*: this operation takes a local TCP address and a remote TCP address as input. It establishes a TCP connection between the two specified addresses.
- *Disconnect*: this operation takes a local TCP address as input. It releases the TCP connection associated with the specified address.
- *Delete TCP end-point*: this operation takes a local TCP address as input. It deletes the TCP port associated with the specified address.

To illustrate the usage of this TLA interface, let us consider the establishment of a point-to-point layer network binding between two CPEs, CPE1(source) and CPE2 (sink), using TCP transport. The sequence of interactions among the connection management components is listed below:

1. FCC to (Internet) LNC: SetupLNB (NFEPPool1, NFEPPool2)

2. LNC creates a LNBM
3. LNC to LNBM: SetupLNB (NFEPPool1, NFEPPool2)
4. LNBM to (TCP) TLA of CPE1: CreateTCPEndPoint (NFEPPool1)
5. TLA of CPE1 creates a TCP port.
6. TLA of CPE1 to LNBM: TCPEndPointCreated (TCPAddress1)
7. LNBM to TLA of CPE2: Accept (NFEPPool2, TCPAddress1)
8. TLA of CPE2 creates a TCP port.
9. TLA of CPE2 to LNBM: TCPEndPointCreated (TCPAddress2)
10. TLA sets CPE2 to accept TCP connection requests from TCPAddress1
11. LNBM to TLA of CPE1: Connect (TCPAddress1, TCPAddress2)
12. CPE1 sets up a TCP connection between TCPAddress1 and TCPAddress2
13. TLA of CPE1 to LNBM: TCPConnectionEstablished
14. LNBM to LNC: LNBCreated
15. LNC to FCC: LNBCreated

In the proposed approach, only the layer network level connection management components are modified to support Internet layer network bindings. Levels below the layer network level are not used for the Internet. This approach makes Internet protocol details transparent to components above the layer network level.

5.8 WRAPPING UP

This chapter has presented a broad overview of NRA. The following characteristics distinguish NRA from other network control and/or management architectures:

- NRA provides network control as well as network management functions. Both classes of functions are designed using a common computational model.
- Both network technology independence and technology integration are central to NRA. By providing a high-level connectivity service and stream bindings, NRA makes technology details transparent to service level applications. For technology integration, the NRA defines components (such as CC and FCC) for managing network connections spanning multiple technologies.
- The computational architecture of NRA for the management functional areas (in particular, connection management) is divided into several levels that make a clear separation between technology-dependent parts (the layer network level and below) and technology-independent ones. This modularity facilitates a smooth introduction of a new technology under the purview of NRA. Further, the modularity enables a network operator to choose different strategies for implementing interworking between TINA systems and non-TINA systems. For example, a net-

work operator may choose not to implement TINA connection management components for layer network level and below for a specific technology, say ATM, but instead use existing systems, such as signaling-based connection setup in ATM networks. For such an interworking scenario, an interworking unit can be designed that provides the LNC interface as defined in NRA but implements the interface using ATM signaling protocols.

- The computational architecture of NRA for the management functional areas is specified in a protocol-independent manner. A network operator has the freedom to choose communication platforms for implementing NRA. While IIOP-based ORB technology is widely believed to be the most suitable platform for TINA DPE, other communication platforms such as CMIP-based management platforms may also be considered by network operators, depending on their business needs.

Chapter 6

Reference points

6.1 INTRODUCTION

This chapter is not for the faint of heart. After reading explanations of the TINA principles in the business model (Chapter 2), service architecture (Chapter 4) and network architecture (Chapter 5) you have now acquired enough TINA knowledge to explore the part of TINA architecture that will probably be the most easily recognizable as *TINA* in any implemented TINA system: *the reference points*.

In the first part of this chapter a brief explanation is given of how the reference point specifications were conceived, while in the second part an overview of each of the reference point specifications is given. These descriptions contain considerable technical detail, and to maintain the character of a textbook, no ODL/IDL descriptions are included. These can be found in the TINA-C documents containing the specifications for each reference point.

A TINA reference point is a point of conformance in a TINA system. It provides the specifications of the interactions taking place between business administrative domains. It is structured according to the TINA viewpoints (see Chapter 2).

6.2 SPECIFYING THE HIGH PRIORITY TINA REFERENCE POINTS

6.2.1 Rationale

To produce a set of specifications that can be used for a wide range of services, the business roles and relationships in Figure 2.2 are mapped on a one-to-one basis onto business administrative domains and reference points, resulting in Figure 6.1. This provides the maximum functionality for each of the reference point specifications. This is also the approach taken in the validation activities in TINA. However, in real-life implementations, the specification can be adjusted to the specific interaction needs of the communicating business administrative domains.

However, due to the fact that the resources in TINA-C are limited, priority was given to specifying particular reference points to facilitate an early and future proof introduction of TINA. The following were considered in selecting the priority for the reference points:

- To promote mass-market acceptance, the reference point between the consumer and the other parties (Ret, TCon and Brk) needs to be specified with high priority. One standard needs to be set to avoid consumer frustration and slow market penetration due to interoperability problems. History provides many examples of this mistake (e.g. the VCR standards: V2000, Beta(max) and VHS) as well as situ-

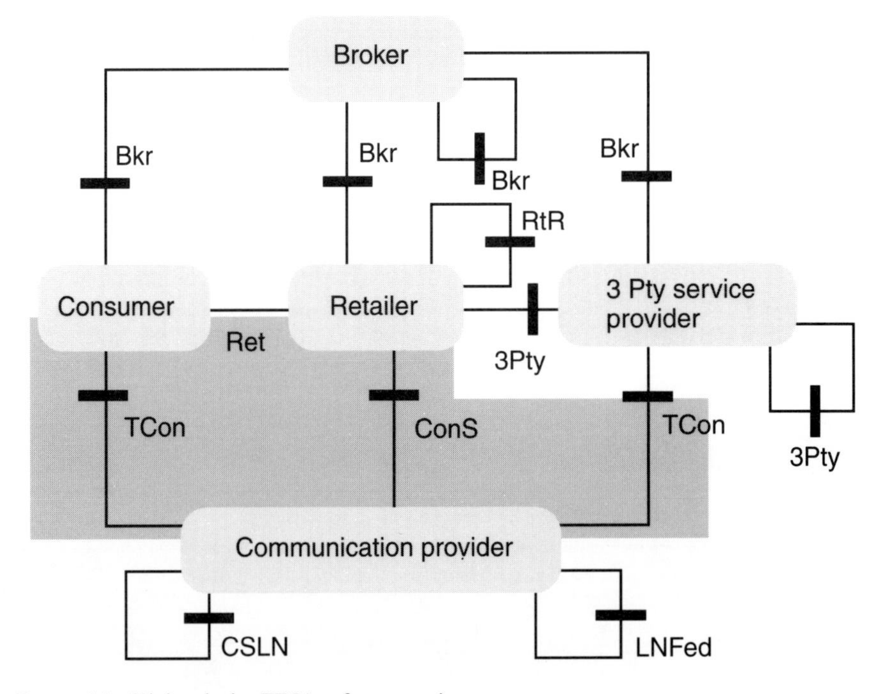

Figure 6.1 High-priority TINA reference points

ations where the industry got it right (e.g. the compact disc standard agreed before introduction of the products).

- To promote the capability to offer a wide range of services by different service providers in TINA, and thus let the market mechanisms (not the technology) decide the services, the consumer in TINA can select from a (potentially large) number of retailers stressing the need for the flexibility of the Ret reference point with its feature sets.
- To locate services and parties in TINA the Brk reference point would be needed. However, these references can be "hardwired" in an initial approach so the Brk reference point is considered lower priority.
- To reduce the investment risk[1] for service providers entering the TINA marketplace, the services need to be decoupled from the (expensive to install) network through the ConS reference point, providing the service providers with a stable communications platform while allowing the flexible introduction of new "service software".

These considerations lead to the prioritization of the specification of the Ret, ConS and TCon reference points.

6.2.2 Combined capabilities of the high-priority RPs

The three high-priority reference points together support the capabilities to set up, control and manage a full range of multiparty, multimedia services (e.g. a multiparty video conference). The Ret-RP provides the capabilities to set up, control and manage a service session between parties. The communication needs of the session are covered though the capabilities of the ConS-RP which allows Flow EndPoint connections to be bound in an arbitrary topology. The TCon-RP supports the access link management in conjunction with the Ret-RP communication capabilities.

To avoid overhead for simpler service session (e.g. a single-event information retrieval service, such as Web browsing) the Ret-RP provides a segment with reduced functionality which allows the user to set up very simple sessions for single-party single-event interactions (the so-called lightweight session).

6.2.3 How the specifications were produced

To provide wide acceptance of these reference point specifications and to benefit from the experience present in the TINA-C community, TINA-C decided to implement a new process, the request for refinements and solutions (RFR/S) process, which could be re-used after the end of the 1993–7 charter of TINA-C.

The following reference points have already been processed in this way and validated by prototyping:

[1] Compare with the monolithic ISDN design and thus the very high cost of introducing ISDN services.

- Retailer (Ret-RP) combining inputs from: Alcatel (on behalf of the ACTS project VITAL), British Telecom, Ericsson, France Telecom, Telia and the TINA-C Core Team described in Chapter 9. A detailed specification can be found in [TC-RR98].
- Connectivity service (ConS-RP) combining inputs from: Alcatel, Ericsson, France Telecom, Telefonica (on behalf of the ACTS project VITAL), the TINA-C core team described in Chapter 9. A detailed specification can be found in [TC–RC98].
- Terminal connectivity (TCon-RP) combining inputs from: Ericsson and the TINA-C core team described in Chapter 9. A detailed specification of the reference point can be found in [TC-RT98].

A high-level explanation of each of these high-priority reference points is included in the following sections.

The following reference points are to be completed in the near future through the RFR/S process:

- Retailer to retailer (RtR-RP)
- Third party (3Pty-RP)
- Layer network federation (LNFed-RP)
- Broker (Brk-RP)

6.3 THE RETAILER REFERENCE POINT

6.3.1 Introduction and scope

The retailer reference point (Ret-RP) defines the interaction between a consumer (playing the user role) and a retailer (playing the provider role). The main function of the Ret-RP is to provide access to end-user services. Additionally, the reference point provides the management of these, as well as the lifecycle management of users (e.g. subscription, administrative interactions).

The Ret-RP, however, does not specify a particular set of services. You will look in vain for the specification of video on demand or teleconferencing in the TINA specifications. The actual service specifications are considered to be much too dynamic and situation-specific for the TINA specification process followed here. The Ret-RP does provide the supporting capabilities (session concept) to support a large variety of services.

The actions provided by the retailer can range from the simple to the very complex, requiring composition of various services from different providers into a new service. The services offered by the retailer through the Ret-RP need not be created in the retailer domain, but can be provided by other business roles: third-party providers, connectivity providers, brokers and even other retailers.

The Ret-RP has been subject to a request for refinement and solution within TINA-C and Version 1.0 is currently being used as a basis for a number of implementations.

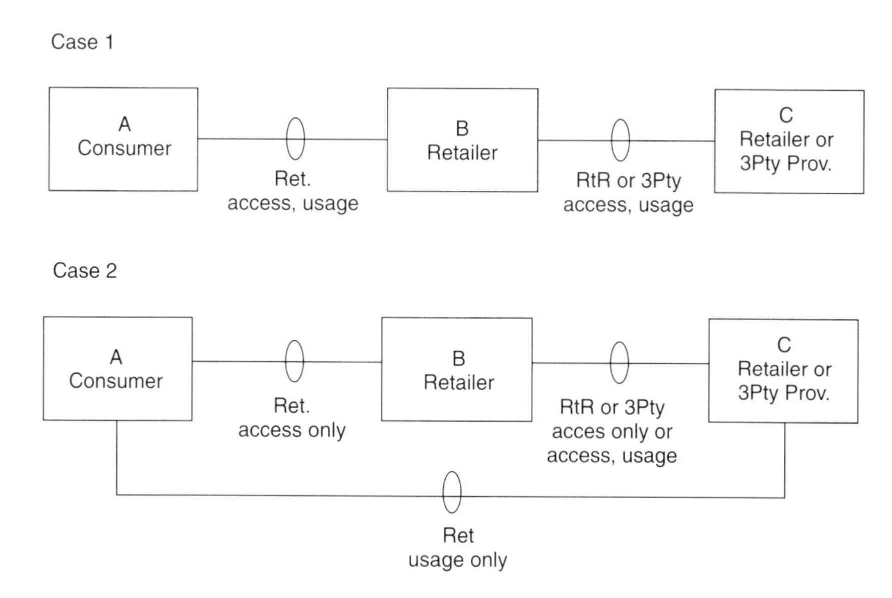

Figure 6.2 Two possible scenarios for the application of the Ret-RP

6.3.1.1 Application of the Ret-RP

Due to the complex relationships that can occur among consumers, retailers and third-party providers, the Ret-RP is not only the most segmented reference point but also supports the concept of delegation. The application of the Ret-RP is shown in two possible situations (see Figure 6.2):

- Case 1, the more straightforward situation: consumer A interacts directly with retailer B, and this interaction takes place on the Ret-RP; on the other hand, interactions between retailer B and a provider C (retailer or a third-party provider) takes place on the RtR or 3Pty reference point.
- Case 2: the consumer performs all access-related interactions across the access segment of the Ret-RP with retailer B. The actual usage of the service, however, requires a direct interaction between consumer A and the retailer or third-party provider C, which occurs via the usage part of the Ret-RP between A and C. Thus, the usage part of Ret must be able to be used objectively in both cases.

6.3.1.2 Business role lifecycle

The Ret-RP provides the interactions to support the complete lifecycle of the relationship between consumer and retailer. It includes the subscriber and end-user lifecycle. The subscriber lifecycle describes the processes by which a subscriber establishes a relationship with a retailer, and modifies or terminates that relationship. The relationship includes subscription, customization, and the association between subscriber and end-user. The end-user lifecycle describes the process by which end-users can access and use services. This includes end-user system setup, retailer contact, and service customization.

6.3.1.3 Segments of usage in the Ret-RP

The Ret-RP can be segmented into two main segments, access and usage. The access segment of the Ret-RP relates to the access session. The usage segment of the Ret-RP is split into the service segment, relating to the service session, and the communication segment, relating to the communication session.

6.3.1.3.1 The access segment

This segment contains three types of interactions:

- The interactions that allow a consumer to initiate a dialogue with a retailer. This contains the interactions to mutually authenticate and set up a secure context and establish an access session. (End-users can remain anonymous.)
- The interaction within an access session to allow a consumer to start and manage service sessions.
- The interactions that allow the retailer to contact the consumer outside an access session, such as to invite the consumer to join a session. To be able to use these capabilities, consumers have to have a persistent domain access interface registered at the retailer to enable the retailer to contact them.

Other reference points will want to use similar capabilities for access related activities (e.g. establishing an access session, starting services, etc.). In order to allow these operations to be re-used in other reference points, a generic access information model and computational interfaces are defined. In the following subsection, consumer can be replaced by user and retailer by provider to create this generalization.

6.3.1.3.2 The usage segments

Due to the complexity of the interactions and the variety of services provided over the Ret-RP, the interactions are grouped into segments, termed feature sets. These can be combined into more complex capabilities and although the segments are specified as stand-alone capabilities, some segments do not make sense without others.

Basic capabilities

The BasicFS (feature set) is the basis for any capabilities provided on the Ret-RP and enables the users to discover interfaces and session models supported by the provider service components. The BasicExtFS extends this basic capability and allows the provider service components to discover the interfaces and session model in the user domain. Currently it is the only session model that is supported in the TINA session model.

Multiparty capabilities
Once the session model is agreed and the interfaces known, the session can be enhanced through the addition of multiparty capabilities from the MultipartyFS and the MultipartyFS with indications (MultipartyIndFS). These allow the participants to retrieve information on other parties, announce the session, invite other parties to join the session and end/suspend participation in the session, and receive indications of changes in other parties' participations.

Normally, the parties in a session will have rights and own information within a session. They can transfer these rights and ownership using the ControlRelationshipFS. The VotingFS enhances this by providing capabilities to vote on the operations to be performed (e.g. instead of the chairperson deciding whether a party can join the session, he or she might want to ask the session parties to vote).

Besides the above there are two feature sets foreseen that will allow sessions to use other sessions to compose services through the CompositionFS and sessions to use special resources (e.g. shared files, video bridges) through the ResourcesFS.

Stream (flow) binding capabilities
The parties in a session have roughly three ways to perform the binding needed to transfer stream data.

1. The parties can request the session to bind the stream interfaces for them, providing only the stream interface references through the participant-oriented stream binding FS (ParticipantSBFS) and with indications ParticipantSBIndFS. The capabilities of the stream interfaces can be discovered previously by using the StreamInterfaceFS.
2. Alternatively, the parties can specify all the individual flows to be set up between them by specifying the flow connections through the SFlowSBFS and the SFlowSBIndFS. In this case the StreamInterfaceFS has to be used by the parties to discover the compatibility between the stream flow EndPoints to be bound.
3. To perform simple single-flow bindings, the SimpleSBFS and SimpleSBIndFS can be used.

Management capabilities
To provide full management of the capabilities above, the next version of the Ret-RP will specify a ManagementContextFS.

6.3.2 Ret-RP Information part

6.3.2.1 Access segment

This subsection describes the types of information passed across the access segment of the Ret RP. It is used as the common vocabulary between the consumer and the retailer while interacting over the Ret-RP. It gives meaning to the parameters passed in the operations on the Ret-RP.

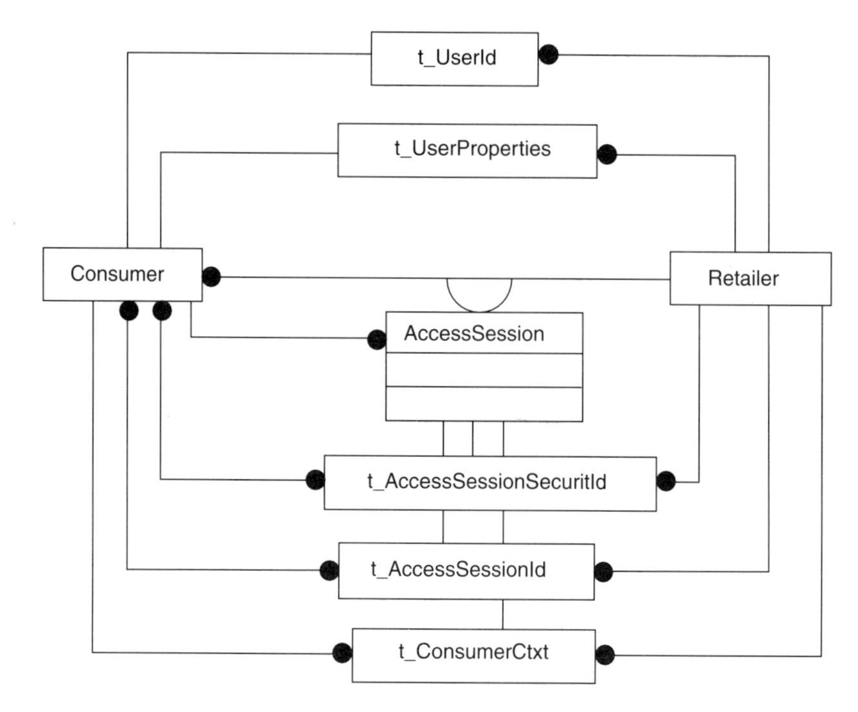

Figure 6.3 Ret-RP access segment information model

Because long-established relationships between consumer and retailer need to be supported, identities (t_UserId), identifying the consumer to the retailer over sessions and property information (t_UserProperties) need to be exchanged. To allow the access session to set the context for the service sessions to, operate the access session needs to be identified (t_AccessSessionId). For security purposes a security context (t_AccessSessionSecurityId) for each session is referenced. To allow the retailer to interrogate the consumer domain on its supported capabilities (i.e. interfaces supported, audio/video processing capabilities) the domain context is defined (t_ConsumerCtxt). See the information model in Figure 6.3.

6.3.2.1.1 Usage segment

In the following, the main information objects for the usage part of a service are given. These information objects allow the parties involved in service sessions to share on vocabulary. In Figure 6.4 the semantics of operations on the session graph are provided. In Figure 6.5 the semantics of the control relationships are provided.

The service session graph (SSG) represents the service session as a whole and contains all information for scheduling the entire service session. A SSG is composed of parties/resources and relationships between parties. A party contains information about a (potential) user and is a negotiating entity that can be an end-user, a subscriber or a ser-

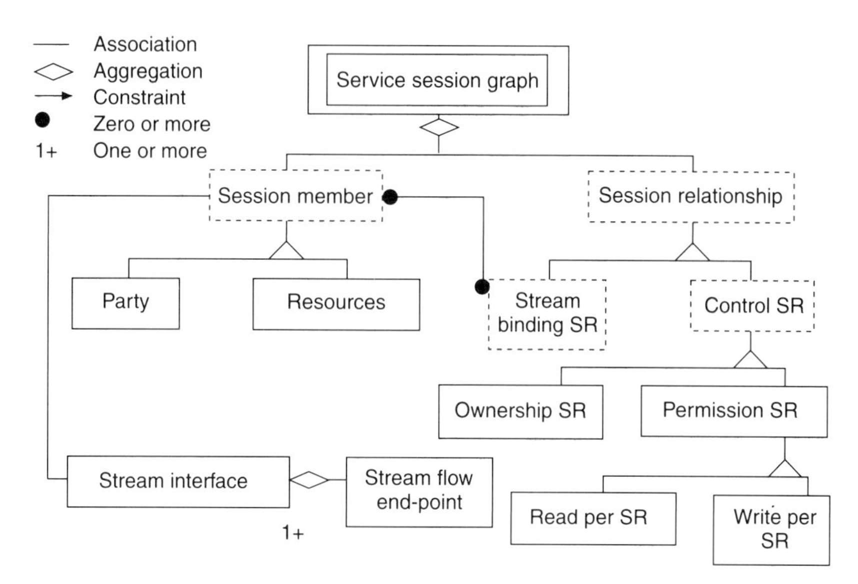

Figure 6.4 The session graph information model

vice or resource provider. If there are already other users involved in the same service session, they may have to confirm the invitation while service session negotiation is taking place. A user can request to remove another one (or him/herself) from the service session (after negotiation) by requesting the corresponding party information object to be deleted.

A resource models a source of support for the execution of the service (session) either shared in the service session by all parties (e.g. a file to be retrieved, a shared pointer in that file (global cursor), a conference bridge, a service or service subscription file, a VoD server, etc.) or a resource private to one of the parties (e.g. background reference library). A resource cannot take part in negotiations. Since parties and resources have many properties in common (e.g. their capability to terminate streams), they are inheriting these from the session member abstract class.

To transfer information between users, the stream flow end-point (SFEP, e.g. a socket of a camera in a video conference session) of these users can be bound by the stream binding session relationship. SFEP can be aggregated into stream interfaces (SI) to allow easier to manipulation, e.g. both video and audio sockets of the camera in the video conference session can be released with one operation. To provide semantics for the actual connection between these stream interfaces, the stream binding session relationship (SBSR, controlled by the stream binding session relationship feature) is introduced. This relationship describes which stream interfaces are bound in which configuration to which other stream interfaces. To allow parties to express the specifics of the session ownership and permissions the information model of Figure 6.5 is used.

The control session relationship (control SR) expresses control capabilities on the service session objects. They express the capability of negotiation and voting and are the basis for the control and voting feature sets. The control SR will always be expressed in

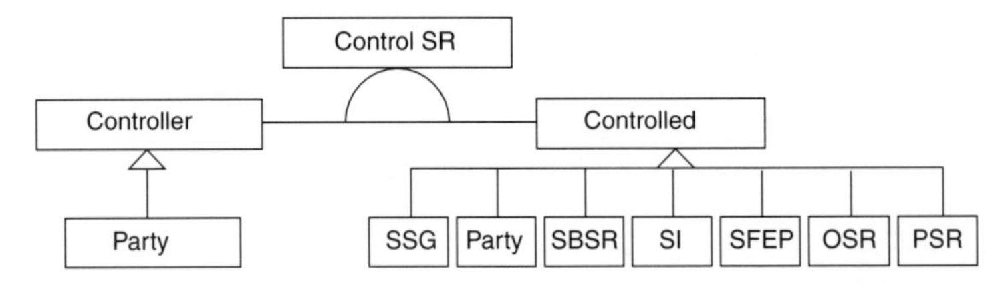

Note:
The 'zero or more' cardinality bullets in OMT do not relate to the describing class for the relation (in this case control SR class), so the correct reading of this diagram is

- A control SR instance must always have one controller and one controlled class instance;
- A controller instance can be related to zero, one or more control SR instances
- A controlled instance can be related to zero, one or more control SR instances

Figure 6.5 The control session relationship information model

terms of a "controller" and a "controlled" role. The controller can control a complete spectrum from the complete service session graph to a single stream flow end-point.

The ownership SR (OSR) specifies which parties have which rights on information objects in the session graph. The ownership includes the rights to authorize the modification if the requesting party is one of the owners or start a negotiation with the owners if the requesting party is not an owner.

Although all this information is used to provide a common vocabulary between the parties in a service session this does not mean that all the information is shared between consumers involved in the session as well. The retailer can provide a "local" view over each of the Ret-RPs between itself and the consumers involved in a session.

6.3.3 Ret-RP computational part

6.3.3.1 The object model

The object model has not been elaborated and is used for illustration only. However, the prescriptive power of the Ret-RP is not impaired. As a consequence the following will deal only with interfaces without grouping them into objects.

6.3.3.2 Objects interfaces for the access segment

Figure 6.6 provides an overview of the computational interfaces exposed in the access segment of the Ret-RP. All the interfaces are categorized according to:

- Which side of the RP offers the interface (consumer, retailer)
- Context previously established (always available, registered to be available outside of an access session, inside an access session).

Registration of interfaces can only be carried out by the consumer on the retailer domain during an access session. The lifetime of registration depends on how the consumer regis-

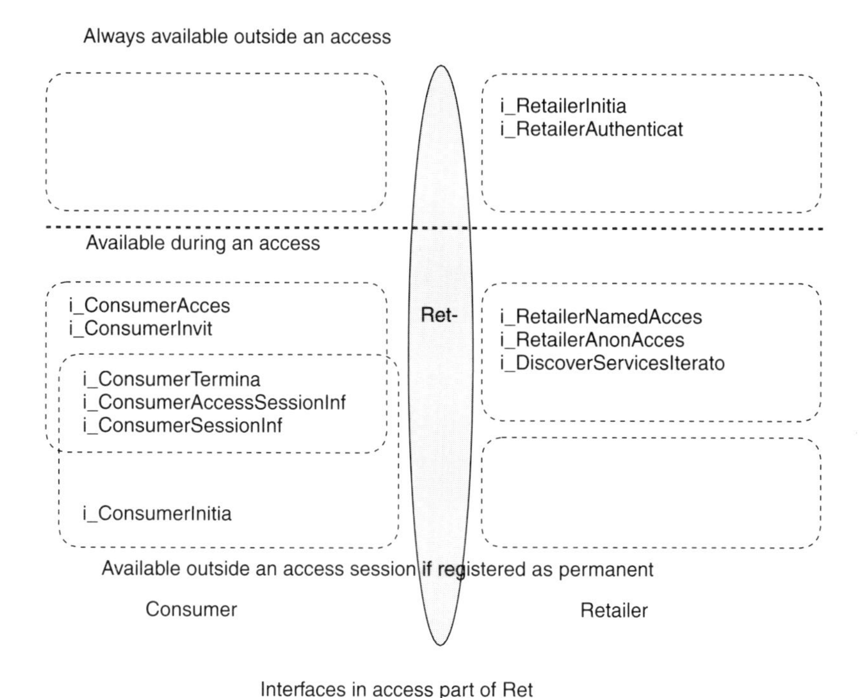

Figure 6.6 Object interfaces for the Ret-RP access segment

ters interfaces, i.e. only as long as the access session exists or permanent. Figure 6.6 names all the interfaces defined by the access part, and categorizes them as above.

6.3.3.2.1 Interfaces always available outside an access session

These interfaces are provided by the retailer to allow the consumer and/or retailer to authenticate themselves, and establish an access session. The i_RetailerAuthenticate interface may be used for authenticating the consumer and/or retailer and passing credentials that can be used to continue within a secure context. The interface i_RetailerInitial checks whether or not the secure context has been set up. If this is not the case, then i_RetailerAuthenticate can be used for doing so.

6.3.3.2.2 Interfaces available during an access session

The retailer supports two interfaces (i_RetailerNamedAccess, i_RetailerAnonAccess) depending on whether a consumer has authenticated as a named or anonymous user[2] to allow the consumer to:

- Access his or her subscribed services

[2] The current definition of the Ret-RP does not allow the change from anonymous to named user in the same access session.

- Start and manage service sessions by returning the reference of the i_RetailerNamedAccess interface
- Register interfaces in the ConsumerCtxt.

The consumer supports interfaces which can be used by the retailer to:

- Find out about the interfaces (i_ConsumerAccess) accessible in the consumer domain and provide the retailer with the interface references (including interfaces outside the Ret-RP, e.g. TCon-RP)
- Notify the consumer of invitations (i_ConsumerInvite) to join service sessions only as long as it has this an active access session. To receive invitations without an access session being active, the consumer must register using the i_ConsumerInitial Interface
- Access terminal configuration information (i_ConsumerTerminal) e.g. applications installed, hardware configuration, (NAPs), etc.
- Inform the consumer of state changes (i_ConsumerAccessSessionInfo) to other access sessions which this consumer has with this retailer
- Inform the consumer of state changes (i_ConsumerSessionInfo) to service sessions which this consumer has with this retailer.

The retailer supports the following interfaces which can be used by the consumer to:

- Give known consumers access (i_RetailerNamedAccess) to his or her subscribed services. The reference to this interface is returned as soon as the consumer has been authenticated by the retailer and an access session has been established
- Allow an anonymous Consumer (i_RetailerAnonAccess) to access the retailer's services. This interface is returned when the consumer calls requestAnonAccess() on the i_RetailerInitial interface
- Discover services matching criteria supplied by the consumer (i_DiscoverServicesIterator) at a given number at a time. This interface can be used iteratively to go through all the services matching the criteria (which can potentially be a large amount if the search criteria are imprecise).

6.3.3.2.3 Interfaces available outside an access session if registered

If the consumer has registered his or her interfaces as permanent the i_ConsumerAccess, i_ConsumerTerminal, i_ConsumerAccessSessionInfo and i_ConsumerSessionInfo are available to the retailer with the same semantics as described above even without an access session. Additionally a new interface is defined that allows the consumer to receive invitations (i_ConsumerInitial) while not in an access session, requesting the consumer to set up an access session to join the session to which he or she is invited.

6.3.3.3 Objects interfaces for the usage segments

In the following a brief description is given of each interface in each segment. Since the diagrams only contain a single interface at a time, no graphical representation is given as with the access segment interfaces.

6.3.3.3.1 BasicFS
This feature set is mandatory for any Ret-RP and needs the TINA session model to operate. The BasicFS provides sufficient functionality to control a single-user session. It allows a client application in the user's domain through the i_ProviderBasicReq interface to:

- Register the client's own interfaces and session models with the session
- Discover the interfaces, session models and feature sets supported by the session
- Retrieve the interfaces supported by the session (both service-specific and those supporting a particular feature set)
- End and suspend the session

6.3.3.3.2 BasicExtFS
This feature set is optional for any Ret-RP and needs the BasicFS to operate. It allows the provider domain session components, through the i_PartyBasicExt interface, to discover interfaces and session models supported by the user domain UAPs. It does not support any session control operations, such as ending or suspending the session, from the provider domain.

6.3.3.3.3 MultipartyFS
This feature set is optional for any Ret-RP and needs the BasicFS to operate. The MultipartyFS allows the session to support multiparty services. It supports requests from the user application for generic multi-party control actions through the i_ProviderMultipartyReq interface (e.g. discovering information about other parties, ending/suspending a party in the session, inviting a party to join a session). It also supports providing information by the session on events that have happened to other participants through the i_PartyMultipartyExe (e.g. announcing the session to the user application, distribution of events when other parties suspend or leave the session, the session asking the user application to be suspended).

The session may use entirely service-specific mechanisms to decide if an action should be performed. Alternatively, ControlRelationshipFS may be used to associate owners to session entities, which determine if an action is allowed. The MultipartyIndFS and VotingFS allow the session to indicate to the user applications that an action has been requested, or to vote on whether an action will be performed. All three may also be used together to determine which parties have rights to perform tasks and vote.

The service session may inform the user domain of changes in the state of the session and its participants through the and i_PartyMultipartyInfo (optional) interface.

6.3.3.3.4 MultipartyIndFS
This feature set is optional for any Ret-RP and needs the MultipartyFS to operate. It allows the session to indicate requests for processing by the user domain applications through the

i_PartyMultipartyInd interface. The MultipartyIndFS allows the session to indicate that an action will be taken shortly (e.g. a user is going to be suspended.). The UAP may be able to vote on whether they wish this action to be taken or not, if the session supports the VotingFS feature set.

6.3.3.3.5 VotingFS
This feature set is optional for any Ret-RP and needs the MultipartyIndFS to operate. The VotingFS supports voting of user applications in a session through the i_ProviderVotingReq interface to determine if a request should be accepted, and executed. The session can supply voting information through the i_PartyVotingInfo interface on the user application. The user application is notified of the impending action by reception of an indication through the MultipartyIndFS.

6.3.3.3.6 ControlSessionRelationshipFS
This feature set is optional for any Ret-RP and needs the BasicFS to operate. It supports parties having, accepting and transferring ownership, and read/write rights on session entities e.g. parties, resources, streams. The interfaces will be specified after the Ret-RP spec is fully completed.

6.3.3.3.7 CompositionFS
This feature set is optional for any Ret-RP and needs the ControlSessionRelationshipFS to operate. It supports the composition and federation of service sessions between domains. The interfaces and operations will be included in the next version of the Ret-RP specification.

6.3.3.3.8 ResourcesFS
This feature set is optional for any Ret-RP and needs the BasicFS to operate. It supports requests and use of resources needed for the provision of a service. The interfaces and operations will be included in the next version of the Ret-RP specification.

6.3.3.3.9 ParticipantSBFS
This feature set is optional for any Ret-RP and needs the BasicFS to operate. It provides the capabilities for the consumer to request the retailer to perform stream binding between a number of participants given the stream interface capabilities of these participants (the i_PaSBFSExe interface). The retailer then invites the other participants (the i_PaSBFSReq interface). Independently of operations being performed the retailer can provide status reports of the status of the stream binding and its parties to the consumers (the i_PaSBFSInfo interface).

It provides the following capabilities for users between members of the session through the i_PaSBFSReq interface on the provider service component:

- Add stream binding to or delete a stream binding from a session
- Add participants to or delete participants from a stream binding: modify the binding topology

- Modify the properties of the stream binding: e.g. the QoS parameters
- (De)activate a previously negotiated stream binding
- (De)activate a previously negotiated party in a stream binding
- List all stream bindings currently active within the session
- Get stream binding properties of a given stream binding

It provides the following capabilities of the user application to the provider service component through the i_PaSBFSExe interface:

- Join or leave a stream binding, indicating that participant wants to join or leave
- Modify participation in a stream binding (e.g. change stream QoS)

It provides the following capabilities of the user application to the provider service component through the i_PaSBFSInfo interface:

- Confirm changes in the status of the stream binding due to previous requests
- Notification of status changes in the stream binding unrelated to this party

6.3.3.3.10 ParticipantSBIndFS

This feature set is optional for any Ret-RP and needs the ParticipantSBFS to operate. It allows the session through the i_PaSBFSInd interface to indicate that an action will be taken shortly: e.g. a stream binding will be deleted. The user application may be able to vote on whether they wish this action to be taken if the session supports the VotingFS.

6.3.3.3.11 StreamInterfaceFS

This feature set is optional for any Ret-RP and needs the BasicFS to operate. This feature set allows session members to exchange stream interface (SI) information as well as associated stream flow end point (SFEP) information. It allows the user to use the following capabilities of the provider service component through the i_SIFSReq interface:

- Query the session on all stream interfaces and their capabilities
- Query the session on the association between the stream interfaces and the participants in the session
- Query the session on all stream flow end-points with particular capabilities and/or belonging to a specific party
- Notify the session that the previously registered stream interfaces are no longer available for use in the session

It allows the user to use the following capabilities of the provider service component through the i_SIFSRegister interface:

- Register one or more stream interfaces and their capabilities

- Modify the registered information
- Withdraw the registered information

It allows the provider service component to use the following capabilities of the user through the i_SIFSExe interface:

- Query the user on the stream interfaces it is willing to expose
- Query the user on the stream flow end-points it is willing to expose
- Notify the user that the previously registered stream interfaces are no longer needed by the session

It allows the provider service component to use the following capabilities of the user through the i_SIFSInfo interface:

- Report the registered stream flows and their stream flow end-points to the user
- Notify the user of changes in the (capabilities) of the registered stream interfaces
- Notify the user of withdrawal of stream interfaces by other users
- Notify the user of the session identifier and stream flow end-point identifiers for this session.

6.3.3.3.12 SFlowSBFS

This feature set is optional for any Ret-RP and needs the StreamInterfaceFS to operate. It provides similar capabilities to the ParticipantSBFS but operates on the individual stream flow connections in a stream binding. The participants in the session do not have to have their stream flow capabilities registered in the session before a stream flow can be set up. This means, however, that the user will need to have knowledge about the characteristics of the various stream flows and their compatibility. It provides the following capabilities for users between members of the session through the i_SFlowSBFSReq interface on the provider service component:

- Add stream binding to or delete a stream binding from a session describing each stream flow connection separately
- Add flow connections to or delete flow connections from a stream binding (modify the binding topology)
- Modify the properties of the stream binding (e.g. the QoS parameters)
- (De)activate flow connections
- List all stream bindings currently active within the session
- Get stream binding properties of a given stream binding

It provides the following capabilities for the users of the session by the provider service component through the i_SFCReq interface:

- Modify (add, delete) a flow connection branch of an existing flow connection
- De(activate) branch flow connections
- Get flow connection information

It provides the following capabilities of the user application to the provider service component through the i_SFlowSBFSInfo interface:

- Confirm changes in the status of the stream flow connections due to previous requests
- Notification of status changes in the stream flow connections unrelated to this party

6.3.3.3.13 SFlowSBIndFS
This feature set is optional for any Ret-RP and needs the SFlowSBFS to operate. It allows the session through the i_SFlowSBFSInd (on flow connections) and i_SFCInd (on branches) interface on the user application to indicate that an action like add/delete or activate will be taken shortly: e.g. a stream flow connection will be deleted. The user application may be able to vote on whether they wish this action to be taken if the session supports the VotingFS.

6.3.3.3.14 SimpleSBFS
This feature set is optional for any Ret-RP and needs the StreamInterfaceFS to operate. It provides capabilities through the i_SimpleSBFSReq interface to set up, modify and delete stream binding where each stream is mapped one-to-one on a stream flow connection. The i_SFCReq interface provides the capabilities to manipulate branches of flow connections similar to i_SFCReq in the SFlowSBFS and pass information back from the provider service component to the user through the i_SimpleSBFSInfo interface. As before, this feature set is dependent on the StreamFlowFS to allow members to acquire SI information.

6.3.3.3.15 SimpleSBIndFS
This feature set is optional for any Ret-RP and needs the SimpleSBFS to operate. It allows the session through the i_SimpleSBFSInd (on flow connections) and i_SFCInd (on branches) interface on the user application to indicate that an action like add/delete or activate will be taken shortly: e.g. a stream flow connection will be deleted. The user application may be able to vote on whether they wish this action to be taken, if the session supports the VotingFS.

6.3.3.3.16 ManagementContextFS
This segment will be defined for the next version of Ret.

6.3.3.4 Event traces for operations over the Ret-RP

The capabilities of the operations described above need to be put in sequence in order to make sense. For this purpose, TINA-C has defined event traces. It would go too far to

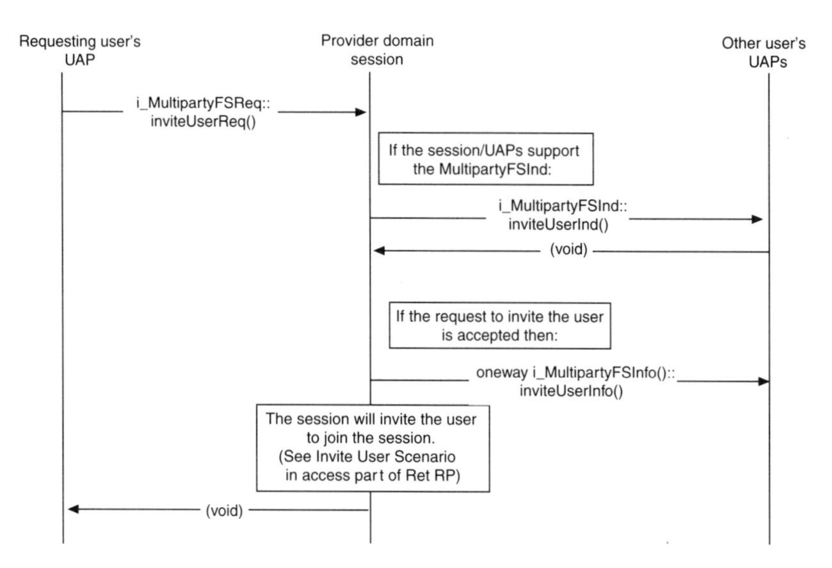

Figure 6.7 Example event trace, invite user

describe all the event traces associated with the Ret-RP in this book, but in Figure 6.7 a typical sequence of operations to perform the invitation of a user into a session from the MultipartyFS is given as an example.

6.4 THE CONNECTIVITY SERVICE REFERENCE POINT (CONS-RP)

6.4.1 Introduction and scope

The connectivity service reference point (ConS-RP) is defined between business administrative domains that provide connectivity services (e.g. leased lines) and business administrative domains that are using those services on behalf of their customers. The connectivity provider may provide two connectivity services to its clients in separate connectivity services sessions.

The first enables connectivity users to set up, modify, and release a connectivity session that is composed of one or more network flow connections. A network flow connection transports information across a connectivity layer network that is made up of one or more layer networks. The end-points of a network flow connection (network flow end-points) may have different characteristic information associated with them: e.g. flow type: ATM, IP. A connectivity layer network can convert types between source and sink network flow end-points (e.g. from ATM to frame-relay).

The second enables a connectivity user to manage (i.e. set up, release, and modify) a group of network flow connections as an aggregated unit (e.g. in a connectivity session, a connectivity user can release all network flow connections that are part of the connectivity session in one operation) making the exchange of information more efficient and thus reduce setup delays.

The generic capabilities of this reference point are:

1. Access segment:

- Authentication of the connectivity user
- Establishment of connectivity sessions for usage segment interactions

2. Usage Segment:

- Connectivity control service:

  Setup, modification, and release of connectivity sessions and network flow connections.

  Event reports on operational state changes of a network flow connection; management of these event reports.

- Contract profile management service:

  Retrieval and modification of contract profile associated with each connectivity user. The contract profile contains default values for parameters exchanged in the usage segment interactions (e.g. which fault reports are notified to the connectivity user) and security parameters exchanged for the Access Segment interactions.

The main interfaces that belong to ConS are grouped into one profile and two main segments (making it a "more simple" reference point than Ret).

For this first issue, it focuses on the connection management aspects. More specifications regarding the other management functional areas will be incorporated into the subsequent version of the ConS reference point.

6.4.2 ConS-RP information part

6.4.2.1 Access segment

Associated with each connectivity user is a contract profile object that contains the context (or profile) information pertaining to a specific connectivity user. This object is created (outside of the ConS-RP) and exists as long as the business relationship between the connectivity user and the connectivity provider exists. The contract profile can be queried and modified via the usage segment of the ConS-RP. For the business relationship between a connectivity user and the connectivity provider to be terminated, the contract profile object associated with the connectivity user must be deleted via the ConS-RP. The authentication information is: degree of trust, authentication protocol to be used, key information, re-authentication interval. The default parameters defined for this version of the reference point in association with the access session are: traffic type, reliability class, and initial administrative state for network flow connections; initial administrative state for connectivity sessions (see Figure 6.8).

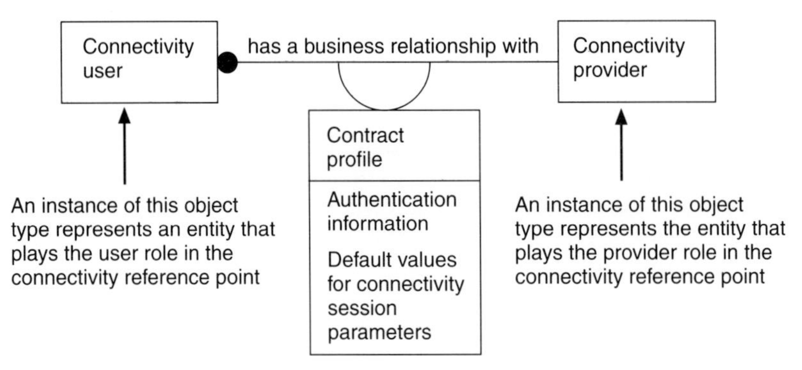

Figure 6.8 Object model diagram for ConS-RP access segment.

6.4.2.2 Usage segment

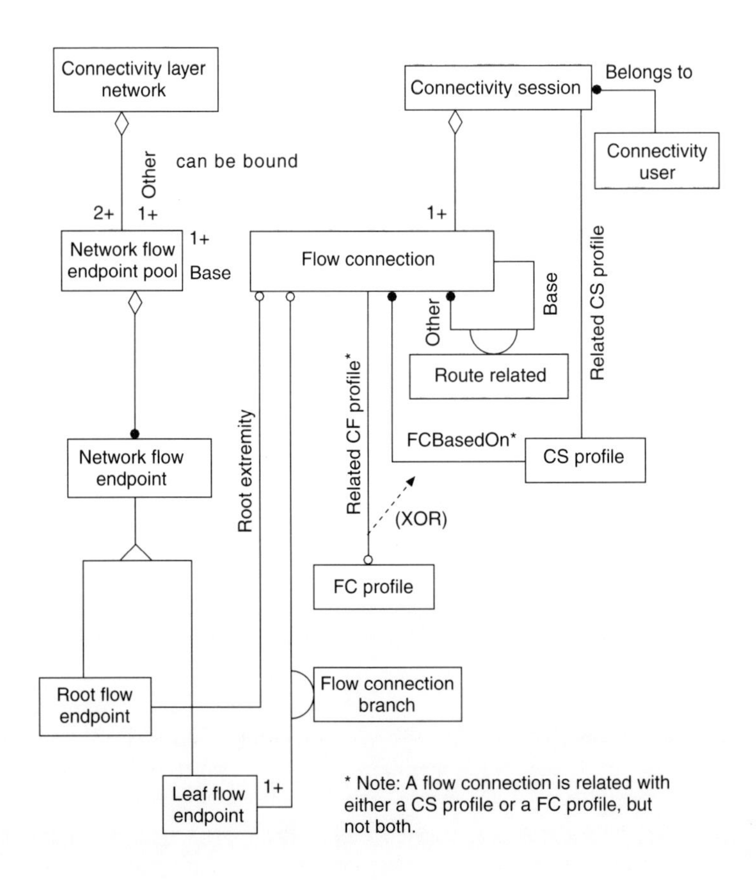

Figure 6.9 OM diagram for ConS-RP usage segment

Figure 6.9 shows the information entities to be manipulated through the ConS reference point. The objects provide the "vocabulary" for both domains to control and manage network connections. In the following, the purpose of these objects is briefly explained, and a broader explanation can be found in Chapter 5.

The connectivity layer network is a container of all the network end-points and is an identifier for the connectivity domain. It represents the "networks" owned by the connectivity provider. It can be used to perform high-level maintenance on a network and check its state. The connectivity layer network contains pools of flow end-points which represent groups of end-points of these network connections. A pool can be the end of a link on which ATM packets are carried with a single wall socket or a group of connections of different technology, e.g. a computer connected to both a telephone network and an Ethernet network using multiple sockets. The end-point is an abstract concept, since it will always be either a root (generating traffic) or a leaf (consuming traffic).

The capability to carry traffic between one root end-point and one or several leaf end-points is labeled as flow connection and allows the identification for manipulation of all the end-points involved in the flow connection. Flow connections can be grouped together into a connectivity session which allows the identification for manipulation of all the connections involved in the session. To express preferences about routing (e.g. to provide lip-synch when video and voice are carried on separate connections through the network) the route related information object is used.

6.4.2.3 State diagram for connections

A connectivity session and its component flow connections have separate administrative states. However, these states are subject to the following rules:

- If the administrative state of a connectivity session is unlocked, then the administrative state of a component flow connection may be either unlocked or locked; it can be individually changed.
- If the administrative state of a connectivity session is locked, the administrative state of each component flow connection is locked.

Similar rules hold between the administrative state of a flow connection and the administrative states of its branches:

- If the administrative state of a flow connection is unlocked, then the administrative state of a branch of the flow connection may be either unlocked or locked; it can be individually changed.
- If the administrative state of a flow connection is locked, the administrative state of each branch of the flow connection is locked.

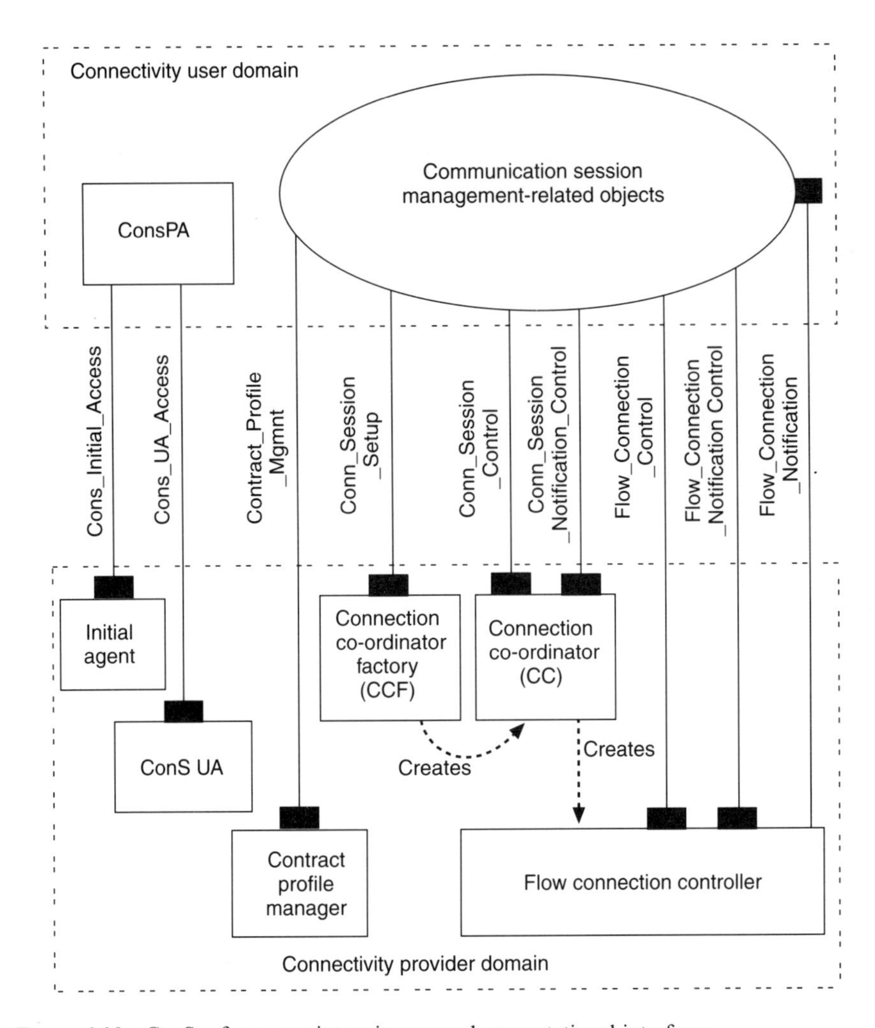

Figure 6.10 ConS reference point main exposed computational interfaces

6.4.3 ConS-RP computational part

6.4.3.1 The object model

Figure 6.10 shows the computational objects exposing their interfaces over the ConS-RP. As explained in Chapter 2 the computational objects are purely shown for explanation purposes, as long as the same set of interfaces (and operations) is offered and the same behavior is exhibited, conformance to the ConS-RP is fulfilled.

For the first version of the ConS-RP, the interfaces are grouped into two segments:

- An access segment of which the (descriptive) objects and their (prescriptive) interfaces are discussed in Chapter 4.
- A usage segment of which the (descriptive) objects and their (prescriptive) interfaces are discussed in Chapter 4.

6.4.3.2 Objects for the access segment

6.4.3.2.1 Initial agent object
This object controls the authentication[3] of the connectivity user through the Cons_Initial_Access Interface. This interface is the persistent generic interface of the connectivity provider business administrative domain. It is registered at the broker, so other domains can request services from the connectivity provider. This interface provides access to the capabilities to discover the access session interface (Cons_UA_Access) provided by a ConsUserAgent object to initiate a specific relationship between the two business administrative domains.

6.4.3.2.2 ConsUserAgent (ConS UA) object
This object represents a specific connectivity user in the connectivity provider domain. It provides the Cons_UA_Access interface exposing the following capabilities:[4]

- Contract profile management service: i.e. retrieval and modification of contract profile information,
- Connectivity control service: i.e. instantiation and control (activation, deactivation, and release) of new connectivity sessions, or control of existing connectivity sessions, establishment of new network flow connections, flow connection branch control (addition, deletion, activation, and deactivation), flow connection control (activation, deactivation, and release), control of notification on changes of the operational state of network flow connections,
- Termination of the business relationship with the connectivity user.

6.4.3.3 Objects for the usage segment

6.4.3.3.1 Contract profile manager object
This object manages the contract profile information associated with a connectivity user. It offers a single interface called Contract_Profile_Mgmnt providing the capabilities to retrieve and modify information of the contract profile information object.

6.4.3.3.2 Connection co-ordinator factory object
This object facilitates the working of the connectivity provider's domain through its Conn_Session_Setup interface by providing the capabilities to:

[3] Since interfaces for authentication services provided by TINA DPE have not yet been specified the principal authentication interface specified in CORBA Security Service Specification [OMG–S] is used.

[4] The actual ODL specifications of the interfaces and operations can be found in [TC-RC96].

- Set up connectivity sessions by instantiation of connection controller objects
- Manage all connectivity sessions in the domain by providing their names on request
- Retrieve references to the Conn_Session_Control and Conn_Session_Notification_Control interfaces of a specific connectivity session.

6.4.3.3.3 Connection co-ordinator object
An instance of this object is created for each connectivity session. It offers the capabilities to:

- Control (set up, activate, deactivate, release) and manage (retrieve session and flow control interface information), flow connections within a connectivity session through its Conn_Session_Control interface
- Control (enable, disable) and management (update of recipient for the notifications) of notifications on the state and state changes of flow connections within a connection session through its Conn_Session_Notification_Control interface.

6.4.3.3.4 Flow connection controller object
An object exists for each flow connection and offers the capabilities to:

- Control (addition, removal, modification, activation and deactivation) of one or more branches of a flow connection as well as management (retrieve information on the flow connection topology, traffic type, routing constraints, etc.) through its Flow_Connection_Control interface,
- Control (enable, disable, set) of the emission of notifications through its Flow_Connection_Notification_Control interface,
- Generation of the notifications which are targeted towards a specific instance of the Flow_Connection_Notification interface in the connectivity user domain.

6.4.3.4 Event traces of operations performed over the ConS-RP

The capabilities of the operations described above need to be put in sequence in order to make sense. For this purpose, TINA-C has defined event traces. It would go too far to describe all the event traces associated with the ConS-RP in this book, but in Figure 6.11 a typical sequence of operations to perform the setup of a connectivity session is given as an example.

6.5 THE TERMINAL CONNECTIVITY REFERENCE POINT

6.5.1 Introduction and scope

The scope of the TCon reference point is the manipulation of network flow end-points

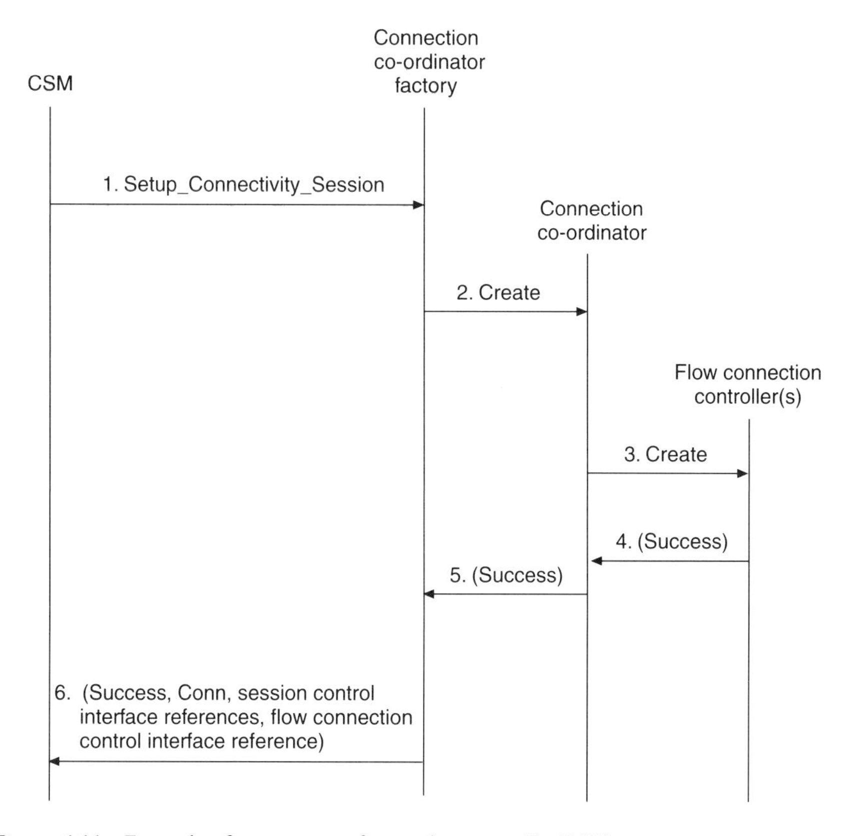

Figure 6.11 Example of a sequence of operations over ConS-RP

(NFEPs) and the network trail termination points (NWTTPs) and network connection termination points (NWCTPs) associated with those NFEPs as shown in Figure 6.12.

The NFEP is the EndPoints of a network flow connection that may span multiple-layer networks. Within the layer network where the network flow connection terminates in the connectivity consumer's domain, the network flow connection is supported by a trail, the end-point of which is the network trail termination point (NWTTP). The NFEP may either be associated directly (bound) to a NWTTP or indirectly by an adapter. On the topological link between connectivity provider and connectivity consumer, the trail is supported by a link connection, the end-point of which is the network connection termination point (NWCTP). The NWCTP is collocated with the NWTTP. The connectivity provider and connectivity consumer are connected by means of a topological link, which has link termination points (LTPs) as its end-points. An LTP contains NWCTPs.

The setup of a NFEP at TCon includes the creation (or selection in the case of pre-provisioned NFEPs) of the NFEP and the creation of the NWTTP. The NWCTP is also created, or in the case of a pre-provisioned link, it is selected. As the NWCTP is identified

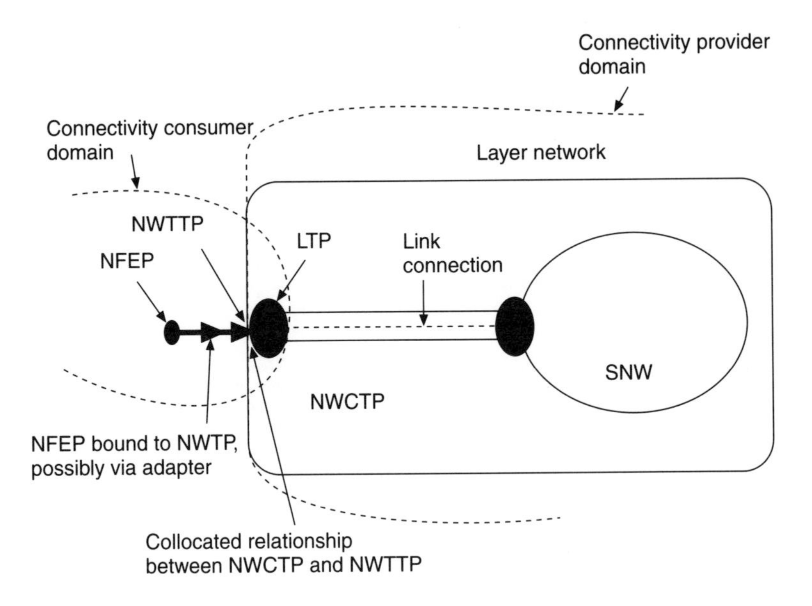

Figure 6.12 Scope of the TCon reference point

by the link identity and the channel on that link, creation or selection of an NWCTP includes channel selection. The concepts shown in Figure 6.12 are explained in more detail in the NRIM TINA baseline [TC-CI97].

The inter-object communication across the TCon reference point may be supported by a DPE. Other ways may be used (e.g. B-ISDN signaling) but these have not been investigated for this version of the TCon reference point.

6.5.2 TCon information part

6.5.2.1 Access segment

The access segment informational objects of the TCon reference point are identical to the one defined for the ConS reference point. The default-values attribute, however, contains one addition: policy information to specify whether the connectivity provider needs configuration change reports and NFEP failure reports.

6.5.2.2 Usage segment

Figure 6.13 shows the information entities to be manipulated through the TCon reference point. The objects provide the "vocabulary" for both domains to control and manage the network flow end-points. In the following, the purpose of these objects is briefly explained.

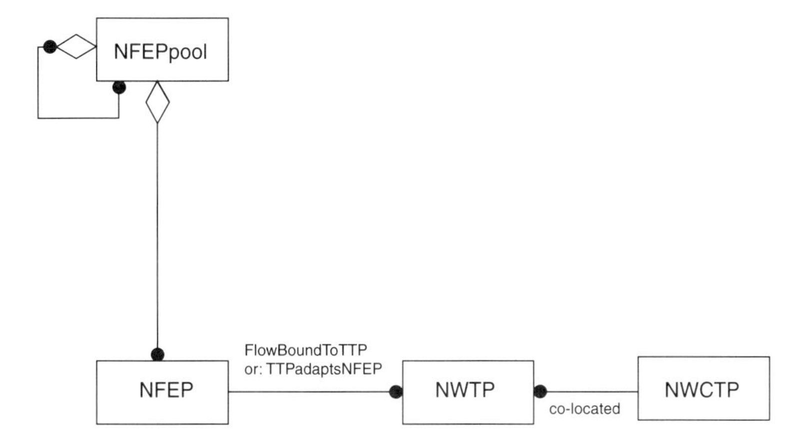

Figure 6.13 Information model for TCon usage segment

Network flow end-point (NFEP) is a technology-independent representation of a particular end-point of a network flow connection. They can be grouped into anetwork flow end point pool (NFEPpool) which itself can be an aggregation of NFEPpools to provide the capability to handle arbitrary groups using simple parameters passed in operations.

A NFEP is FlowBoundToTTP to a network trail termination point (NWTTP) which can be co-located with a network connection termination point (NWCTP) (see NRIM [TC-CI97]) providing the means to relate trails, connections and the respective end-points in the network. TTPadaptsNFEP implies that an adapter is used between NFEP and NWTTP to convert from one technology to another: e.g. a TCP/IP protocol stack to terminate an ATM link.

6.5.2.3 State diagram for NFEPs

Figure 6.14 shows the state diagram for a NFEP. The state diagram indicates which transitions can take place in the administrative state of the network flow end-points.

Once created, a NFEP can be in two states:

- Locked: In this state the NFEP is fully configured, but is not ready to send and/or receive information.
- Unlocked: In this state the NFEP is fully configured, and is sending and/or receiving information.

Creation, deletion, and transitions between the states are achieved by operations on the computational object that encapsulates the NFEP information object, and is described in Chapter 5.

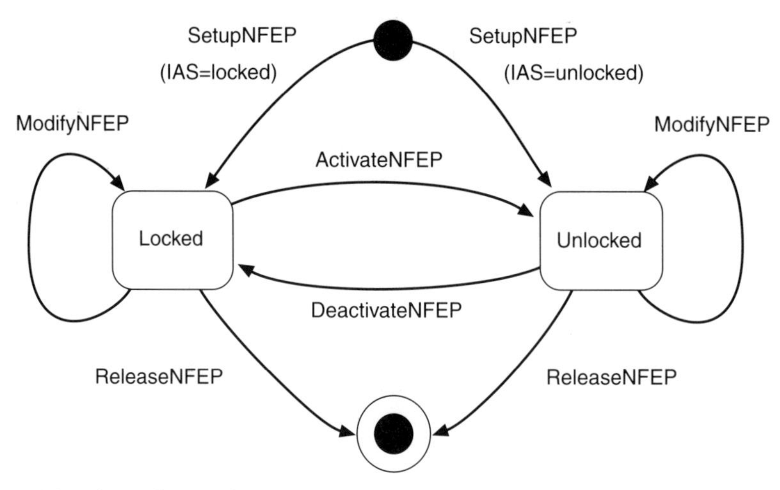

Figure 6.14 State diagram for a NFEP

6.5.3 TCon computational part

6.5.3.1 The object model

Figure 6.15 shows the computational model for TCon. TCon prescribes the interfaces shown in this figure and the behavior of the connectivity consumer and connectivity provider with respect to those interfaces. The individual objects are shown for ease of understanding, but are not prescriptive. An implementor may define a distinct set of objects. As long as the same set of interfaces is offered across the TCon reference point, and the same behavior is exhibited by the overall set of interfaces, compliance with the TCon reference point is fulfilled.

As with the previous reference points, the interfaces on TCon are categorized into an access and a usage part. The usage part itself is subdivided into "Control of NFEPs", and "Configuration management".

6.5.3.2 Objects for the access segment

6.5.3.2.1 TCon provider agent (TConPA) object

The TConPA represents a connectivity provider in a connectivity consumer's domain. An instance of a TConPA is created every time access interactions are needed, and lasts at least as long as these interactions. It provides the following capabilities:

- Start access (invite) with the connectivity consumer by setting up a secure binding through the i_tconpaInitial interface
- Context management (get_reference) for location of connectivity consumer domain interface references by the TConUA to through the i_tconpaCtxt interface.

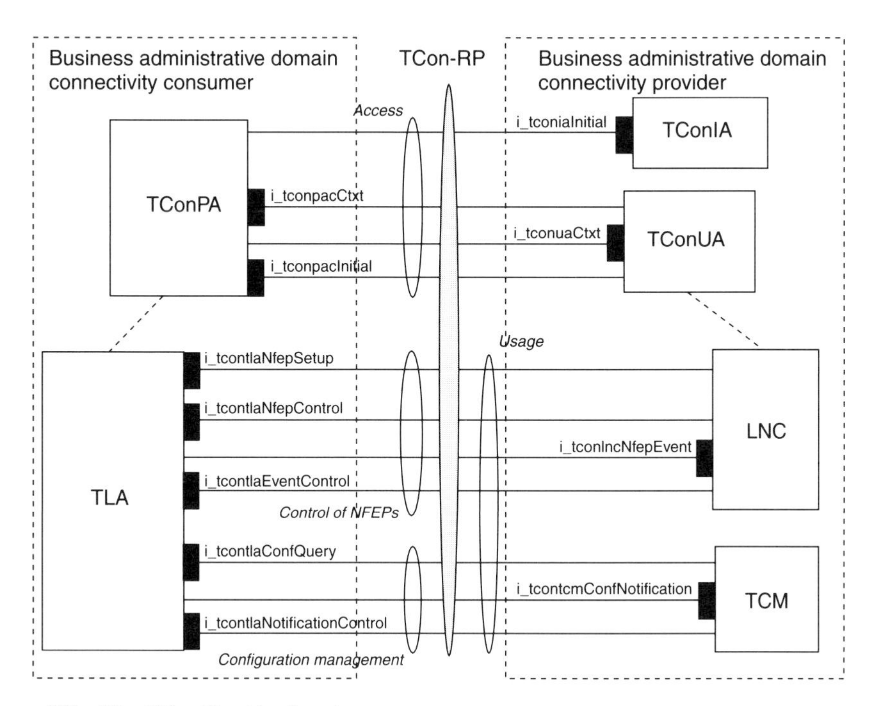

TConPA = TCon Provider Agent
TConUA = TCon User Agent
TLA = Terminal Layer Adapter
LNC = Layer Network Co-ordinator
TCM = TCon Configuration Manager

Figure 6.15 Computational model for TCon

6.5.3.2.2 TCon initial agent (TConIA) object

This object is the initial access point to a connectivity provider's domain. It provides the capability to request the start of access interactions to the connectivity provider by obtaining an interface reference to its user agent through the i_tconiaInitial interface. The reference to this interface can be supplied either by electronic means (e.g. registered in a trader) or by off-line means (paper procedure).

6.5.3.2.3 TCon user agent (TConUA) object

This object represents a connectivity consumer in a connectivity provider's domain. There is one instance per connectivity consumer of a TConUA in a connectivity provider's domain. This object is created when the business relationship between connectivity consumer and connectivity provider is set up, and exists as long as the business relationship exists. It provides the following capabilities:

- Context management, related to domain access security (get and update authentication information), related to the network flow end-point management (get and update default NFEP values), related to domain policy information exchange (get and update) as well as the retrieval of usage interfaces references in the connectivity provider domain through the i_tconuaCtxt Interface

6.5.3.3 Objects for the usage segment

The specification of the TLA, LNC and TCM and their interfaces, as provided by this document, is technology independent. In some cases the given specification may directly be usable for implementation, but specialization for different technologies is allowed for other cases.

6.5.3.3.1 Terminal layer adapter (TLA) object

This object provides functions for control (creation, manipulation, deletion) of network flow end-points (NFEPs). There is one TLA in the connectivity consumer's domain for each type of layer network. It offers the following capabilities:

- Setup of NFEPs in the connectivity consumer domain through the i_tcontlaNfepSetup interface. When the operation is performed successfully, the NFEP will exist, and be bound to a NWTTP and its co-located NWCTP.
- Manipulation (activate, deactivate, release (see the state diagram of Figure 6.14) and modify NFEP characteristics) of NFEPs through the i_tcontlaNfepControl interface so it can be used by the connectivity user.
- Control (enable, disable, set notification destination) the emission of event notifications on the state of NFEPs to the LNC through the i_tcontlaEventControl interface.
- Query (get_nfeppools) the TLA about names of available NFEPpools and their status through the i_tcontlaConfQuery interface.
- Control (enable, disable, set notification destination) the emission of notifications by the TCSM regarding available NFEPpools through the i_tcontlaNotificationControl interface.

6.5.3.3.2 Layer network co-ordinator (LNC) object

The LNC manages trails and tandem connections within a layer network domain. It interacts with the TLA to manipulate the end-points of these connections. There is one LNC in the connectivity provider's domain for every layer network for every administrative domain within the connectivity provider. The LNC controls NFEPs by performing operations on the TLA's i_tcontlaNfepControl interface and allows/accepts notifications from the TLA of state changes of NFEPs through its i_tconlncNfepEvent interface.

6.5.3.3.3 TCon configuration manager (TCM) object

This object maintains a view on the available connectivity configuration in the connectivity consumer's domain. To this end it may actively query the i_tcontlaConfQuery interface

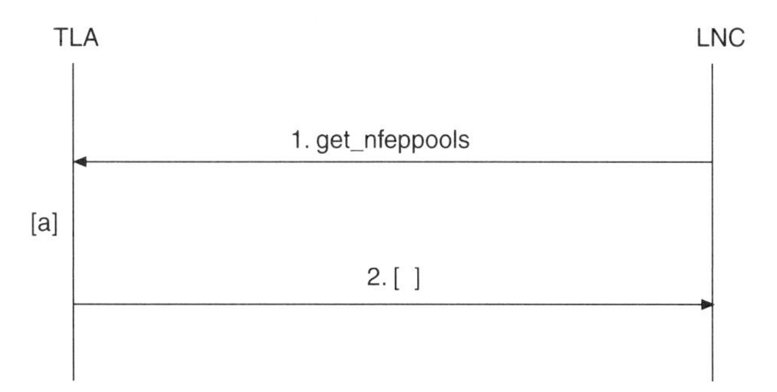

Figure 6.16 Example of a sequence of operations over TCon-RP

on the TLA object (see Chapter 5) in the connectivity consumer's domain, or it may be notified of configuration changes by the TLA through its own i_tcontcmConfNotification interface (nfeppool_created, nfeppool_deleted).

6.5.3.4 Event traces of operations performed over the TCon-RP

The capabilities of the operations described above need to be put in sequence in order to make sense. For this purpose, TINA-C has defined event traces. It would go too far to describe all the event traces associated with the TCon-RP in this book. However, in Figure 6.16 a typical sequence of operations to perform negotiation of event notification is given as an example.

Most of the interactions sequences over the TCon reference point look like the above, since they are mostly single atomic operations. This makes the TCon reference inherently simpler to implement than Ret and ConS.

6.6 INTERDEPENDENCE BETWEEN THE HIGH-PRIORITY REFERENCE POINTS

All the high-priority reference points do not require any of the other TINA reference points to be implemented. However, in case other TINA reference points are also implemented the following inter-reference point relations apply.

6.6.1 Interdependence between ConS and TCon

The scope of ConS is the management of network flow connections. The end-points of these network flow connections are managed through TCon. Manipulations of network flow connections at ConS will require/result in manipulation of network flow end-points at TCon. For example, the setup of a network flow connection at ConS will result in the setup of a network flow end-point at TCon.

6.6.2 Interdependence between Ret and TCon

The Ret reference point carries information on the binding between a stream flow end-point (SFEP) and a network flow end-point (NFEP) within the consumer domain. This will result in management actions on TCon for the NFEPs. To interrelate the actions on TCon and Ret a special correlation identifier is introduced.

Chapter

7 Integration of TINA into existing networks

The installed base of the telecommunication networks represents one of the largest global investments since the beginning of the industrial age. In 1997, the public switched telephone networks (PSTNs) alone represents around 800 million lines worldwide with an annual growth rate of about 3-4%. The total annual investments exceeded US$150 billion in 1997.

The ongoing transition towards the information age will generate another surge of investments in data and computing networks. Today's Internet growth is a good indicator of this trend. The installed base of Internet users, hosts and network elements (300 million in 1998) generates an aggregate rate of about 20 Gbit/s compared to 1000 Gbit/s in the voice transit telephony network (PSTN) today. In view of the phenomenal growth rate of Internet services, the total aggregate Internet traffic is expected to equal the PSTN traffic by 2001.

Looking for new business opportunities it is also worth while to note that the notion of legacy systems itself is blurred, and the part of the existing infrastructure that is considered to be in business for a time frame of some 20 years is decreasing. In contrast, the average time period for renewing capital investments has been reduced to less than a decade, sometimes even as short as two years. That's why the reality of today's telecommunications world is made of heterogeneous environments, with different transport and data networks, a variety of access means and an increasing number of terminal types, as well as – of course – a tighter and tighter coupling with the computer world.

Introducing complete TINA flexibility into existing networks implies the introduction of new interfaces. Because the modification of public networks requires major investment and a long lead time to market, it is necessary to use migration paths based on legacy interface and signaling for a rapid introduction of TINA. Intermediate migration steps will most probably keep legacy signaling interfaces (INAP, for instance); this means that the present problems such as the lack of "clean-cut functional separation" may remain for a while.

Newcomers without "heavy" legacy systems might see that as an advantage, e.g. over big telcos whose interests is to protect large past investments and to provide a smooth transition from old to new rather than a green-meadow approach. Those integration efforts might slow down the traditional telcos, on the one hand, but, on the other, they can build on a dominant market presence and a solid customer base.

Today's existing different technologies do not necessarily converge towards a homogeneous environment. This generates the need for interoperation at several levels, and also for common solutions across multiple networks and platforms. At the network level, each transport system is – so to say – "autonomous" in that it guarantees a specific access or delivery service: cellular phone systems grant mobile telephony, IP and the Internet the transport of information packets with "best effort" characteristics, and so on. On the service level the problem exists that in most cases each existing network comes bundled with its own specific services on top, most likely unable to interoperate. One example is voice mailbox services offered by cellular operators for their users being completely separated from voice mailboxes on the Internet, even if the service is the same. To overcome this potential lack of integration capability this chapter focuses on explaining how TINA systems can be plugged into existing or near-to-be telecommunications networks, thus adding value mainly in the service layer, aiding as a service platform for integration of different services and networks.

Two business opportunities are briefly recalled here, as they could actually be the ones that trigger the integration of TINA into existing networks. The first is the growing necessity to provide seamless services across networks operated by different stakeholders. The second is the necessity for a telecommunications operator, who may wish to satisfy the constantly growing demand of complex and customizable services over different networks, to turn its monolithic and heterogeneous network control/management systems into a unified, simplified, open and programmable software infrastructure. Both opportunities can be summarized in providing multi-service integration built on a multi-network integration.

7.1 TINA AS MULTI-SERVICE INTEGRATOR BUILT ON MULTI-NETWORK INTEGRATION

First, it is worth spending some time on what is actually meant by integration, as compared to migration or interworking. In the following, a TINA system is considered as either (1) a separate "island" system bridged to the "mainland" network systems through interworking units (IWU), (2) a system introduced into a particular network system as the result of an upgrade of software components, or (3) a subsystem as such in an existing system. The first case is referred to as an interworking scenario, the second as a migration path, the third as an integration of a TINA system into an existing network. It is intended in this chapter to put more emphasis on integration, although the frontier is obviously blurred and, to a certain extent, not crucial.

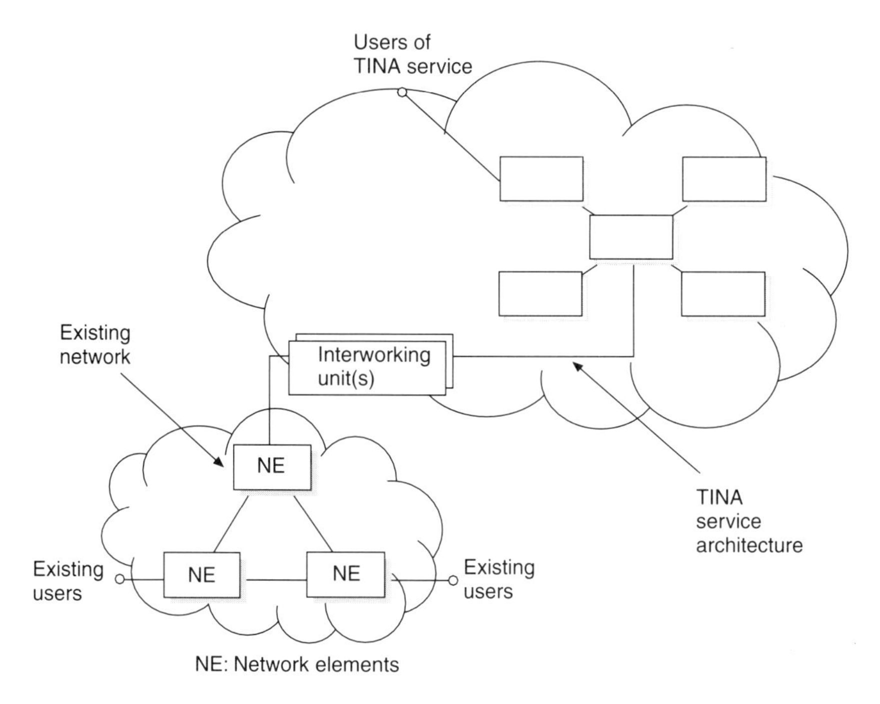

Figure 7.1 General interworking concept

7.1.1 Protection of investments in the installed network base

In order to become successful in the market, TINA products must not change or destabilize the existing network elements. The strategy of choice is to provide an interworking concept which will enable the existing networks, network elements and services to interact with TINA solutions for carefully selected features. No changes in existing network elements should be required. This will avoid network instability problems, revenue loss and customer complaints. This interworking requires adapters or specialized components. Figure 7.1 schematically illustrates this principle. Existing network can be voice, where TINA services for instance are voice over IP.

The aim is to incorporate the interworking unit(s) into the TINA architecture in order to minimize impact on existing networks and network elements. The scenario in Figure 7.1 enables users to benefit from some part of the TINA service (limited by, for example, the capabilities of the existing user's terminals).

Since today's networks and their features are not static but subject to a dynamic growth, a prerequisite for TINA success in the marketplace is its capability to deal with this phenomenon. For example, if a new service usage accounting feature is to be introduced into an existing network, the TINA service should be able to support this new feature without disturbing the existing service feature set. Figure 7.2 illustrates this principle:

The new TINA features match the extensions in the existing network and should

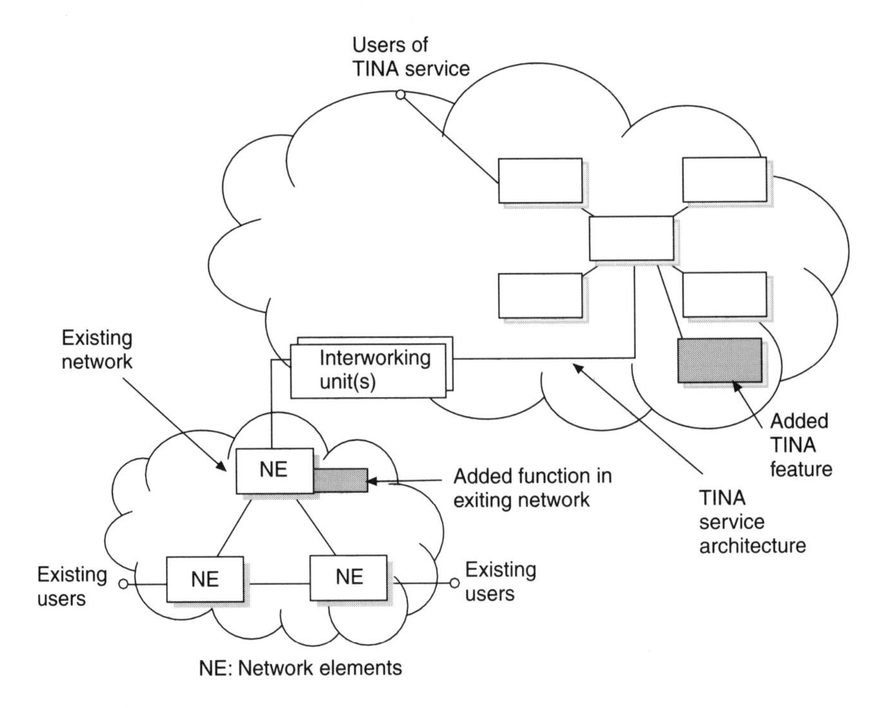

Figure 7.2 General migration strategy

eventually be replaced by a "pure" TINA solution. More and more new features can gradually be implemented in the TINA context. In the above example, the new service accounting feature (shaded boxes in Figure 7.2), can be fully implemented as a part of the TINA service architecture. This again assures existing network stability without impeding feature growth. With its distributed service principle the TINA architecture is particularly friendly to this migration strategy.

A word of caution is, however, necessary. No one can predict network features in the future. It is the market that dictates them. This means that the TINA architecture must be open towards new features, interfaces and network technologies. In this sense TINA represents a standardization philosophy different from those of established standardization bodies (e.g. ITU-T, ETSI, ANSI). TINA-C does not aim at defining a detailed interface or feature standard, as is the case with the traditional signaling protocols (CCS7, etc.), but rather it is an extensible reference architecture with many degrees of freedom with respect to adding new interfaces, reference points and functional blocks.

The interworking concept and migration strategy are not specific to TINA. These concepts were the principal forces behind any innovation in the telecommunication networks in the past and will continue to play a major role in the future. There are numerous examples: starting with the introduction of digital technology for PSTN, introduction of advanced features for intelligent networks (IN), network management (TMN), growth of corporate and data networks, etc. They all adhered to the above rules for interworking and migration.

7.1.2 Telecommunications services on heterogeneous networks

The TINA service architecture serves as a common platform for providing advanced services to a variety of interacting telecommunications networks. Its powerful session concept promotes the TINA service architecture as the common service layer for a heterogeneous telecommunications environment. The access session (based on components like the user agent) concerns a specific user/provider relationship: the user is associated by means of specific objects to profile, subscription, settings and preferences information, independently of the networks and terminals being used for information delivery. In this way, applications have direct access to the whole information characterizing the user allowing supporting fully personalized services. This is somehow different from current situations in which user-related information and profiles are scattered over different network elements (e.g. in the local switch, in the service control point, in the home location register (HLR), and so on).

As an example the case of accessing to unified messaging systems could be considered. A person can access his or her voice mailbox from a mobile handset, computer, touch-tone telephone, etc.; he or she will probably like to access to a *single* mailbox, ubiquitously accessible, provided by a network operator, a retailer, a "voice mailbox service provider" or any of these actors. This is possible with current systems, but it requires *ad-hoc* agreements among providers (opening up of interfaces that sometimes can be proprietary) and *ad-hoc* implementations for supporting interworking of protocols and systems are to be put in place. Conversely, this unified access is possible with the TINA service architecture, because a user will simply set up an access session with a retailer from any terminal, access his or her profile, start the "voice mailbox" service (whether provided by the retailer itself or by any third party) and use it by exploiting distributed processing mechanisms and well-defined and open interfaces. Note that, even if the user terminal does not have a DPE, alternative mechanisms can be established for the user-to-retailer interaction, while the DPE enables all interactions from the retailer to the whole telecommunications infrastructure.

Likewise, the service session concept enables multiple users attached to a variety of networks to share a common context, that is, to use a service together. Even more, various users could hold an individual perspective of a service (in terms of QoS, service features, privileges, etc.) while sharing information and resources with the other parties in the session; or various parties in a session could use different services individually, but still communicate and share some service content.

Figure 7.3 illustrates this situation. Users access the service platform via a variety of networks; each can be targeted to a specific class of customers, and each single customer can also access the service platform from different networks: cellular phone users via the mobile access network; "classic" telephone users via the plain switched telephone network (PSTN) or directly via ISDN, both relying on the intelligent network for access to the service layer; users of the Internet and of the World Wide Web, with a PC or a network computer; businesses having their computer networks connected directly to IP and VOIP, or to

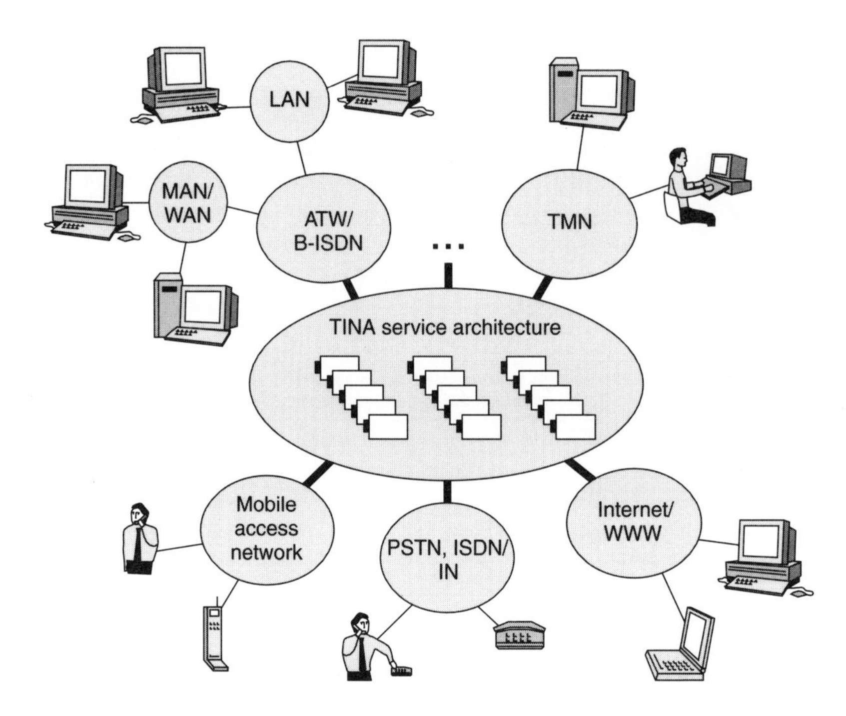

Figure 7.3 The TINA service architecture as a common service layer for multiple networks.

an ATM backbone, controlling connectivity via B-ISDN signaling; management operators of private and public networks, requiring increasingly advanced management services, and many more. They all share a common service platform and, in many cases, common services. This requires the service platform to support a variety of protocols and interfaces in order to interact with several network infrastructures. In this sense the service platform structured à la TINA has to *adapt* to the control and management interfaces of the underlying networks. Two ways to build the adaptation level and to make the TINA service architecture the "point of control" for a number of different networks do exist: the *adaptation unit* approach, which means inserting an extra component that acts as gateway between TINA and each underlying network, and the *components specialization* approach, which means tailoring some key components in the service architecture to a specific underlying network; both are discussed in the following sections.

7.1.3 Adaptation unit approach

The adaptation unit approach relies on the principle that the service architecture components are fully independent of the transport network being used. A specific component (the adaptation unit) receives protocol messages that are specific of a network protocol and refer to a non-TINA reference point (e.g. INAP for IN, UNI for B-ISDN, HTML for the

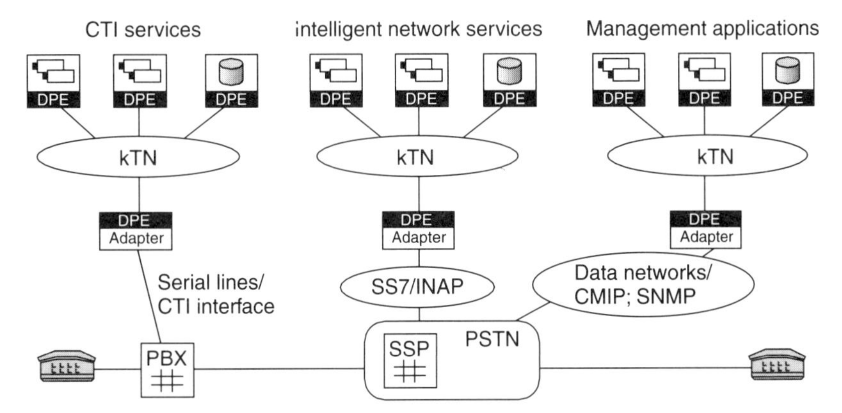

Figure 7.4 TINA integration into candidate telecommunications networks

WWW, etc.) and converts them into network-independent invocations on IDL interfaces offered by TINA components, and vice versa.

The network systems which can contribute to this unified, open, programmable and multi-network software infrastructure comprise the customer premises equipment, including corporate processing nodes integrating the private exchange resources (CTI), intelligent public networks, the telecommunications management networks (TMN), the Internet the Voice Over IP, and digital audio-visual networks and retailer systems (DAVIC). The adaptation unit approach is depicted in Figure 7.4.

In this approach the TINA service architecture components remain unchanged. Of course, service implementations must take into account the constraints imposed by the network (for example, in terms of transport capabilities, e.g., bandwidth) and by the non-TINA interfaces (for examplc, in tcrms of limitation of the network protocols and related network functionalities, e.g. support for third-party connection/call control).

The obvious advantage of this approach is that the service architecture in itself (without adaptation units) is exactly the one defined by the TINA Consortium and can be considered network-independent. The drawback is that the adaptation units can be extremely cumbersome in terms of design and implementation, and can become a significant bottleneck.

7.1.3.1 Specialization of components approach

In the specialization approach, service components are specified (inheriting by TINA object classes) and implemented with an underlying network in mind. They provide the functionality as specified in TINA, or as needed for a given service, but they are specialized – in the sense of the explanation in section 4.4 – for a specific non-TINA reference point.

In other words, the adaptation functions are not separated from the TINA functions in that TINA components offer IDL interfaces that support standard protocols and APIs as defined for a specific network. The advantage is that the total complexity of the system is reduced, as studies demonstrate [EURESCOM Project P508], because from general-purpose components we shift to more *ad-hoc* components. The adaptation functions are more distributed, so the risk of bottlenecks is considerably reduced. TINA solutions can be offered for supporting and re-using legacy systems.

The drawback is that some TINA functionalities are replicated and engineered more than once. It may become difficult to keep track of protocol versions, emerging systems, etc. in a large number of components, instead of few adapters. Note that this approach still enables the service architecture to be a common infrastructure for multiple networks; *ad-hoc* components share not only a common paradigm (object-oriented DPE-based communication, session models, etc.) but also common information (profiles, etc.).

In both approaches, it is crucial to develop a TINA "componentware", that is, a library of TINA components, with variable degrees of service independence, that can be purchased in a multi-vendor market. In this way service designers can create a large number of services for a large number of networks and at the same time take advantage of an advanced service platform.

To understand how TINA systems as a whole will be plugged into existing networks a more detailed view on integration aspects is described below, considering not only telecommunication applications object-based software components from the service layer. Additionally the kernel transport network, interconnecting a variety of computing nodes, and a kernel middleware, based on an object request broker and some system components, which is deployed on that nodes, will be investigated.

Among telecommunications applications, a distinction is made in TINA between network resource management applications (connection set-up, fault management, configuration inventory, etc.) and end-user services (plain old telephony services, information retrieval services, communication services, etc.). To illustrate their integration a closer look will be taken at connection management.

Following the same lines examination of how to integrate TINA into existing networks is made first, from the kernel transport network viewpoint, second, from the middleware viewpoint, third, from the connection management architecture viewpoint, then from the service viewpoint. Finally an example of integrating different existing networks using WWW and IN services under the umbrella of the TINA service architecture and a migration path from GSM to UMTS, to serve fixed and mobile integration will show the benefits of the generic nature of TINA systems.

Obviously such a global multi-network software infrastructure will not appear in one day – a phased introduction of TINA into these systems is necessary. The phasing the most referred to, from the most "TINArizable" telecommunications networks to the least, consists of integrating TINA into network management systems first, since they are already object-oriented and they do not have stringent (e.g. real-time) constraints. It will most likely be followed in VOIP networks, since no control and management systems actually exist, then in intelligent networks, in ATM networks or in corporate CTI systems, in order to provide higher quality of services. Within the short term TINA will be introduced into the wireless networks, for which a substitution by an open and object-oriented infrastructure cannot be envisaged in the short term, given the investment currently made.

7.2 TINA IN SMALL OFFICE/HOME OFFICE TERMINALS

As often recalled, the main target of the TINA Consortium is to produce an open, object-oriented software architecture for telecommunications networks operated by telecommunications companies and third-party software systems. The focus of the Consortium is therefore on network intelligence or network computing. Nevertheless, a requirement upon which the member companies largely agree is that this next generation intelligent software architecture must not impose a totally new, telecommunications-specific software technology for the customer premises equipment, e.g. laptops, personal computers, set-top boxes, and so forth, to be software-wise pluggable into telecommunications networks. What is required is a network computing architecture that can be prolonged seamlessly, with no particular or additional investment for the customers, towards these end-point software pieces. The objective is reached, as explained below.

In the case of end-user terminals running DPE-like middleware, like simple World Wide Web browsers, browsers with IIOP applets, applications encapsulating Active X controllers, Java-based software, MHEG engines, etc., the invocation of TINA services deployed in the network causes no major problems. This invocation can occur first because the client of a network object request broker (ORB) and of its computing servers does not have to run a fully fledged ORB (as long as the terminal is not offering services to the active network). Second, it can occur because interworking issues between IIOP-based ORB, on the one hand, and Microsoft COM, HTTP and Java, on the other, have been worked out by IT providers for quite a while and are solved by now.

In the case of less intelligent terminals, terminals like phone sets which can only interpret a small set of user-to-network signaling features (for example, the invocation of the TINA services full set) cannot occur directly. Special network resources like interactive voice systems or intelligent peripherals, capable of extending the user-to-network dialog to the level expected by the network computing services, can be inserted between the less intelligent terminals and the network computing servers.

Another alternative, also provided by the TINA service features, is to authorize different "capability sets" for the TINA customer-to-retailer reference point, also corresponding to the better-known user-to-network signaling interface: a capability set for dumb terminals (phone sets) based on traditional signaling protocols, a capability set for more intelligent terminals based on Internet-oriented protocols, and a dynamically negotiated capability set for richer end-user computing environments with downloading capacities, for which the customer-to-retailer reference point or user-to-network interface has no fixed semantics.

7.3 TINA IN DIGITAL AUDIO-VISUAL NETWORKS (MULTIMEDIA)

The objective of the Digital Audio-VIsual Council (DAVIC) was to offer integrated digital networking for services like telephony, video-telephony, video-conferencing and high-speed data interchange, the resulting network, based on the ATM technology, and software infrastructure being intended for use by residential, home office and small business ser-

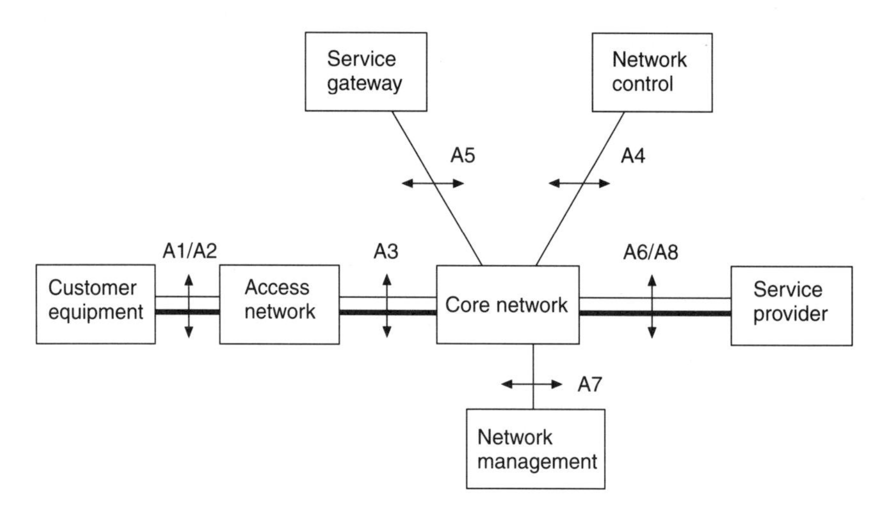

Figure 7.5 The DAVIC reference model

vices. The physical reference architecture retained by DAVIC was characterized by access networks capable of transporting information flows over ATM channels, a core transport network based on ATM over a SDH transmission infrastructure, and (Level 1) service gateways connected to a variety of service providers (Level 2 GateWays – L2GW). The customer premises equipments (CPE), residential or SOHO, are intended to handle MPEG-2 encoded information streams.

Among the set of reference points identified in the DAVIC 1.1 Reference Model and depicted in Figure 7.5 some are more relevant to an integration of a TINA system, like (A6) between the digital audio-visual servers and the DAVIC network infrastructure, (A1) between the customer premises equipment and the DAVIC network infrastructure, (A4) between the core network and the network control system, (A5) between the service gateway and the network control system, etc.

7.3.1 Integration of a GIOP-based kernel network into digital audio-visual networks

DAVIC adopted the Internet Inter-ORB protocol standard as a communication support for the DSM-CC messages. The software infrastructures worked out by TINA-C and DAVIC are aligned from this viewpoint.

7.3.2 Integration of an OMG middleware into digital audio-visual networks

DAVIC specified the semantics carried out by the DSM-CC protocol by means of the OMG Interface Definition Language. Although the use of OMG-compliant middleware is not required by DAVIC, such a platform provides developers of digital audio-visual applications with development tools and system components that are undeniably suitable and easy to use.

7.3.3 Integration of TINA connection management services into digital audio-visual networks

In the DAVIC reference model, the service gateway truly acts as a retailer system, which interprets customer demand for a particular service provider and orders the establishment of a connection between the equipment of the requesting customer and the server of the selected service provider. The Council has not imposed the format of the request for end-to-end connections between customer equipment and information servers.

7.3.4 Integration of TINA end-user services into digital audio-visual networks

The design of a DAVIC service gateway following the TINA service architecture principles would clearly benefit the DAVIC system, since no detailed specification of the service gateway has been proposed or worked out so far. As outlined by [TINAGate] the functionalities to be supported by a service gateway range from communication session control to resource management and brokerage.

7.4 TINA ON THE INTERNET AND VOICE OVER INTERNET

For brevity, the Internet network can be seen as a "plain vanilla" transport network based on the IP protocol, at the periphery of which are interconnected end-user terminals with IP or PPP/SLIP protocol stacks, a large number of application servers offering HTTP, FTP, H323, Gopher services and a small number of Internet access providers (IAPs). Internet software components, which can range from FTP virtual terminals to sophisticated HTML browsers with integrated applets, are therefore essentially at the periphery of the Internet network, in the terminals – servers and user terminals – not in the network itself. Nevertheless the Internet has some disadvantages.

First, the Internet network is stateless (no sessions): packet routing is not predictable, routers have no knowledge of previous routings, there is no concept of virtual circuit, and no user context containing, for example, user identification, except in H323. The various HTTP requests are stored and managed as a kind of access session that provides the user with continuity, confidence and customization.

Second, the quality of service of the transport network cannot be guaranteed and negotiated. The Internet network is best-effort: no bandwidth can be reserved, no throughput or response time guaranteed.

Third, the Internet protocols suit information-retrieval applications, such as retrieval of HTML pages. They suit transactional applications less; a more object-oriented protocol (IIOP) is currently proposed to suit the latter.

Therefore the Internet also needs to evolve and it will, undoubtedly, as initiatives like VOIP and Internet 2 confirm. Are the TINA architectural concepts good candidates for the technical directions that the Internet should take? The answer is thought to be positive, for the main reason that IT companies leading the Internet software market are already pushing technical directions that will be aligned with those made in TINA.

7.4.1 Integration of a GIOP-based kernel network into the Internet

The integration into the Internet of a kernel network based on the GIOP communication stack is occurring independently of the work carried out on the TINA software architecture by the TINA-C members. This integration first started with the appearance of SunSoft Java applets or Microsoft ActiveX controllers in the terminals being able to send IIOP or COM requests over the network.

7.4.2 Integration of an OMG middleware into the Internet

The integration of a distribution transparent middleware like the one specified in the OMG in the Internet computing nodes (the user terminals and the information servers) is also occurring, independently of the TINA-C effort, under initiatives like the collaboration of IBM, Netscape, Sun and Oracle to join forces and work on open standards for network computing. Such collaborations should enable interoperability between applications created from a variety of development tools, components and platforms like IBM's Visual Age, Oracle's Network Computing Architecture (NCA), Sun's WebServer products and Netscape's Open Network Environment.

The purpose of this work carried out by information technology key players is even to transcend an Internet-versus-object standard battle, so that customers and developers can make their programs in ActiveX/COM, CORBA and HTTP/CGI work together seamlessly. This obviously paves the way for a rather facilitated introduction of TINA components in the Internet software domain, and very precisely in the customer terminals, since no telecom-specific middleware will be required in order to run code interfaced with the proposed network services.

7.4.3 Integration of TINA connection management services into Internet

Internet is based on the Internet Protocol IP, i.e. datagrams. With IP, no connection is established at the network level between the customer terminals and the information servers, whereas logical connections are established at the TCP level. Many contributions can be found on comparisons between connection-oriented (e.g. ATM) and connectionless (IP) networks, and the purpose of this section is not to open the debate here. Thus, it is assumed that the network provider looks for a solution capable of guaranteeing customers a quality for IP services, such as an information throughput guaranteed independently on the network load, a satisfying end-to-end transport delay and an upper limit for jitter.

Although the possibility of abstracting the quality of service provided by the two types of IP transport network (ATM or RSVP based) at specific computational (i.e. programming) interfaces of flow control component has not been proven, such an abstraction would transcend the ATM-versus-IPv6 battle to the benefit of all. The TINA connection management architecture has been designed originally for connection-oriented networks, i.e. networks transporting data flows after establishing the end-to-end route for the stream containing these data flows. Since the TINA components controlling the data flows and streams are, on the one hand, hierarchical (each handles stream control at a specific network layer level) and, on the other, distributable on the various computing nodes that make the network control plan, it is very likely that TINA represents a frame capable of abstracting protocols for reserving flows and protocols for establishing connections.

7.4.4 Integration of TINA service components into the Internet

The Internet is already a global network with widely used services like the World Wide Web, electronic mail and file transfer; it constitutes a powerful means for access and usage of information. Internet solves the issue of interoperability, but not that of the "application organization". Furthermore, it relies on a very specific network view: all intelligent processing is handled at the periphery of a transport network which is requested only to transmit data flows, anonymous bytes in opaque pipes, and not to process them.

The integration of TINA end-user services into the Internet will undoubtedly ease TINA system implementations and at the same time enhance some Internet services. One domain in which such integration is first likely to occur is that of Voice Over IP, the domain precisely at the frontier between the traditional telecommunication companies world and the Internet. Integrating TINA service-oriented concepts or components into an Internet access provider (IAP) and/or ISP system would enable context-oriented access sessions instead of stateless and volatile HTTP sessions to facilitate service integration like controlling telephony connections from a third-party WWW server, to open up the reference points of the IAP systems in order to provide roaming between the various points of presence, etc.

What can the TINA service architecture bring to this fast-evolving Internet world? It provides a way to construct services and applications: using the TINA "philosophy" (distributed processing, re-usable components, etc.) to design Internet applications surely brings advantages. In the specific case of the Internet, the service architecture, and in particular the session concept, enables the provisioning of *integrated information and telecommunications services* using the Internet and the telecom infrastructure for transport, the World Wide Web for access and TINA for service applications, including the seamless post of legacy telephony services.

As mentioned earlier, TINA has to play the role of a service platform for integration of different networks. Taking this approach, an IP network can be seen as another transport infrastructure to be used by TINA services for supporting the communication needs of users. Developing an IP application using TINA service architecture makes it portable to ATM and SDH when needed. A good example is CSCW or video-conferencing.

7.5 TINA AND CTI SYSTEMS

A noticeable effort has been made in recent years to standardize the communication protocol between the corporate telephony systems (PBX) and the computer systems. This has been undertaken by different organizations like ECMA working on the standard "Computer-Supported Telecommunications Applications (CSTA)", ANSI T1 on the standard "Switch Computer Application Interface" and ITU-T specifying TASC; "Telecommunications Applications for Switches and Computers". Two types of specifications have emerged from these standards: specifications of communication protocols between telephony equipment and a computer and specifications of the software services supported by computers. At the same time, application programming interfaces have been defined and provided to application developers by software companies like Microsoft (TAPI), Novell (TSAPI) and SunSoft (JTAPI).

The recent work undertaken by alliances like Versit, created in 1994 by Apple, AT&T, IBM and Siemens, or the Enterprise Computer Telephony Forum, created by DEC, Dialogic, Ericsson, Hewlett-Packard, Motorola and Northern Telecom, intended to merge the specifications of the protocol between telephony equipment and a computer and the application programming interface encapsulating the computer-supported telephony services, is very comparable to the TINA architecture specification work mainly intended for public telecommunications applications. Since the work is proceeding along similar lines and the business models used by these organizations are complementary, there is much to gain in working on the interoperability of the types of architecture.

7.5.1 Integration of a GIOP-based kernel network into the CSTA

Computer-supported telephony applications (CSTA) are mostly provided in a distributed environment, based on local area networks and usual communication protocols. In CTI applications, GIOP and DCOM are already used between servers and terminals, e.g. telephony application ALM of Novell, using IIOP.

7.5.2 Integration of TINA connection management services into the CSTA

CSTA can be used to control connections from a TINA connection management. Thus using TINA service architecture above TAPI/CSTA makes application portable from private networks (PABX) onto public networks. This will be particularly useful for applications like call centers and added value on PABX used by new operators in their public networks.

7.6 TINA IN INTELLIGENT NETWORKS

As one of the cornerstones of the development of TINA, interworking and integration with IN is a primary target for its application. TINA can implement service needs, where IN falls short (e.g. perform complex information exchange between users and the network). In IN a clear separation is introduced between the delivery segment (handling of connections) and the network intelligence (control of the connections and services). An overview of the IN functional architecture is given in Figure 7.6.

The connection-handling segment is implemented using the CCAF and CCF functionality and the connection and call control segment is implemented using the SST, SCF, SRF and SDF functionality. When implementing an intelligent network, the CCF and SSF and sometimes the SRF are packaged into the exchange (or service switching point SCP) while the SDF, SCF and sometimes the SRF are packaged into a service control point (SCP). When the SRF is maintained as a separate entity this is called the intelligent peri-

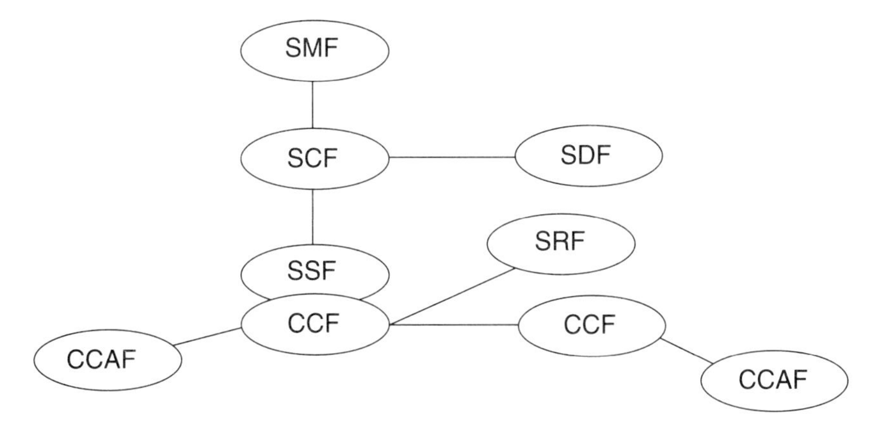

Figure 7.6 The intelligent network functional plane

pheral (IP). The exchange and the control point communicate through the INAP protocol implemented as an application part of a C7 protocol stack.

Studies have been made by the EURESCOM project P.508 "Evolution, Migration Paths and Interworking to TINA" in order to assess the different scenarios of migration, and interworking of IN systems towards, and with, TINA systems. The possible integration or migration scenarios leading to a fully open programmable service platform are given below. The integration of TINA solutions into IN networks can be envisaged as follows (see also Figure 7.8 as a refinement of Figure 7.2):

1. Replacement of the Signaling System 7 (SS7) transfer capabilities by the TINA kernel transport network based, for example, on the Internet Inter-ORB Protocol (IIOP) and TCP/IP
2. Replacement of the INAP protocol stack and operating systems in the SCP, IP and SSP with the TINA ORB platform to run the intelligent services. This will allow the distribution transparencies of the ORB to be used and promotes the re-use of "IN" components
3. Implementation of TINA service object-based components related to connection setup and management, in replacement of, or in addition to, the connection control system provided in intelligent networks
4. Implementation of TINA service object-based components related to end-user services, in replacement of, or in addition to, the intelligent service independent building blocks.

7.6.1 Integration of a GIOP-based kernel network into IN

The replacement of the current Signaling System 7 (SS7) kernel networks based on the transaction capability application part (TCAP), the signaling connection control part

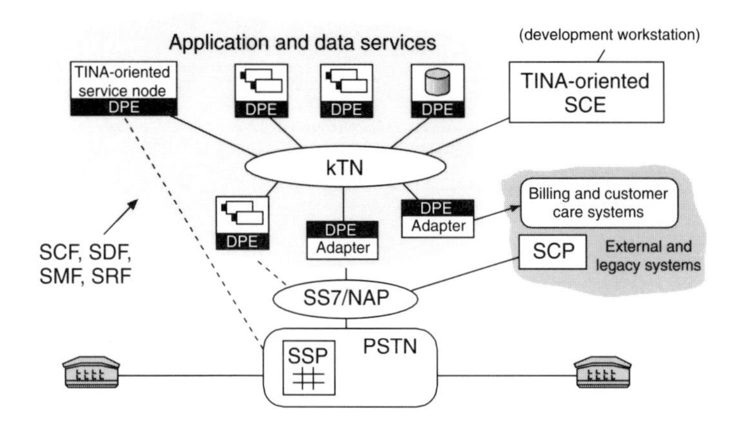

Figure 7.7 A SCP distributed through a DPE

(SSCP) and the message transfer protocol (MTP) by a IIOP/TCP/IP-based kernel network has the advantage of reducing the number of communication protocols used between the processing nodes of the intelligent networks and of facilitating the deployment of a CORBA-based middleware. The integration of such an IIOP kernel network can be envisaged at different physical points in an intelligent network, one most likely being the SCPs (Figure 7.7).

The legacy interworking is done by an adapter unit which matches TCAP/SSCP/MTP protocol stacks to IIOP/TCP/IP stacks. Computationally, this unit can operate in two ways:

1. Translate the existing remote operation service element (ROSE) specifications of the TCAP messages into IDL specifications and vice versa. The translation implies that the intelligent network application part (INAP) is interpreted quite late in the chain, by the CORBA object offering the interface covering the TCAP semantics as shown in Figure 7.8.
2. Through object adapters, which can directly route the incoming INAP request to the pertinent stub for unmarshalling. In this case, TINA service components do not see the INAP interface directly, but through special components that receive messages on IDL interfaces on top of a DPE and convert them into the INAP/SS7 protocol. This excludes direct use of INAP/SS7 directly by TINA objects, which could limit the application (Figure 7.9).

7.6.2 Integration of an OMA middleware into IN

The replacement of the current middleware used in the IN SCP and the service management points – centralized operating systems like Unix or VMS with different communication protocols, INAP, SNMP/CMIP, DAP – by a distribution transparent OMA-based

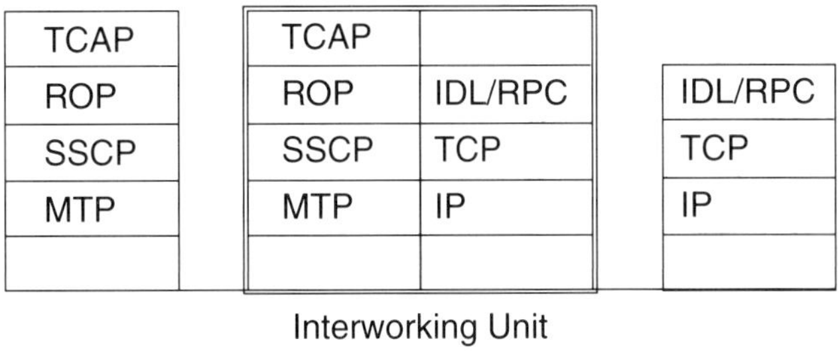

	Interworking Unit		
TCAP	TCAP		
ROP	ROP	IDL/RPC	IDL/RPC
SSCP	SSCP	TCP	TCP
MTP	MTP	IP	IP

Figure 7.8 TCAP/IDL adaptation unit

middleware can bring great advantages in terms of intelligent service deployment and system functionalities. Indeed, since the service control (external) processors are becoming increasingly numerous in order to provide telecommunications services like cordless telephone mobility (CTM), customer advanced mobility enhanced logic (CAMEL), plain switched telephony services, deploying a distribution-transparent middleware on these various processors enables a consistent application programming interface (API) for the communication and computing resources made available on each node.

7.6.3 Integration of TINA connection management services into IN

The SSF models the basic call state model (BCSM) advocated in the capability set 1 of the IN standard. This model is intended for single-medium, two-party calls on fixed networks. In parallel, proposals and studies are made to go further and to modify this basic call state model to exploit the possibility for a separation between call and connection controls. This separation at the BCSM level is one possible step towards integrating TINA connection setup components (here the low-level ones like the TINA connection performers) in replacement of the BCSM connection control components. Such integration would not necessarily consist of modifying the signaling protocols (DSS 1, DSS 2, ISUP, B-ISUP)

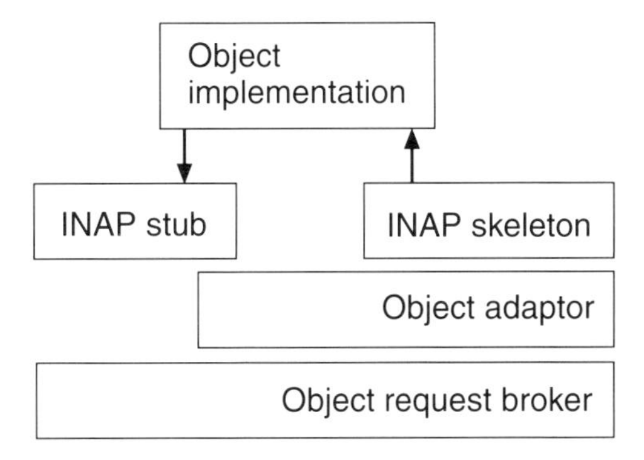

Figure 7.9 Specific object adaptors and stubs for TCAP/INAP

currently used to exchange messages between the switches, but in re-engineering their control software units, as a first step.

Another option is the introduction of the TINA connection setup higher-level components (communication session managers, connection coordinators, layer network coordinators) on top of the BCSM connection control components of the exchanges (separated out or not from the call control ones) and the current signaling protocols. Two physical architectures emerge from the different scenarios of deployment of these TINA connection components: a terminal-initiated connection control scenario first-party connection control initiated by the end-user and a third-party connection control, initiated by the service control.

The end-user connection control scenario consists of deploying in the customer premises equipment (as Java applets or Control ActiveX) some high-abstraction TINA connection management components (e.g. a communication session manager) encapsulating the user-network interface provided by the network to which the equipment is connected. The application programming interface offered by these components deployed in the equipment looks very much like most of the first-party connection control APIs (e.g. WinSock API or Microsoft TAPI).

The network-initiated connection control scenario, on the other hand, consists of deploying some or all TINA connection management components on the external and distributed computing nodes bound to the network equipment. For the service designer, this scenario does not result in a single application programming interface like the Novell Telephony Services API (TSAPI) but in a set of distributed object interfaces that can be invoked in a much more flexible way. Currently, first implementations of this use of TINA are offered in so-called service nodes.

7.6.4 Integration of TINA end-user service components into intelligent networks

The integration of TINA end-user services components is likely to consist first of introducing the access session-related components in the intelligent network service control function, so that access to the intelligent network service portfolio opens up and is no longer bound to a particular access network technology (e.g. plain-old telephony network with basic user-to-network signaling), and so that the richness of the TINA access service model (customer context durability, access customization, access security) benefits also the current intelligent services, especially with CS1 in local exchanges.

This integration must be made consistent with the recommendation currently worked out in ITU-T Study Group 11 on user-to-SCP signaling (USS), since obviously both aim at providing direct dialogs between users and the service control functions, without the interference or the screen of the connection-control-oriented basic call state model. The TINA access session model constitutes a global model for the various "out channel call related user interactions".

7.6.5 Example integration of TINA, WWW and IN services

Intelligent network services integrating the Web capabilities are attracting increasing interest. Several proposals (e.g. see [http://etsi.fr/tiphon]) are emerging in this field. These

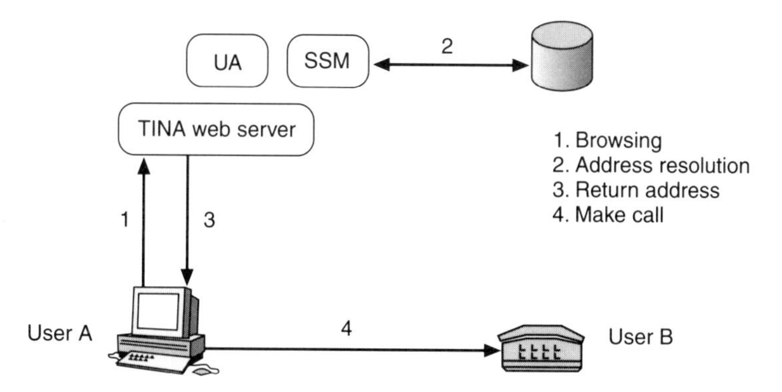

Figure 7.10 A TINA web server resolving telephone addresses

services are based on a general scenario that allows the introduction of several TINA solutions. This class of services is based on the ability to browse the Web and, on demand, to establish connections (making a telephone call or by means of internet telephony) through the browser to a user.

A first option (Figure 7.10) considers a Web Server (hiding a TINA infrastructure) able to provide users with a service similar to White Pages, e.g. a list of freephone numbers. The assumption here is that the originating terminal is able to treat, in an integrated way, IP- and telephone-based communication (e.g. using computer telephony integration applications). The service is provided according to the following phases: user A searches for the telephone number of user B, the TINA Web server gives the telephone number of B, the user A makes a telephone call to B.

The second option (Figure 7.11) considers a Web server able to control by means of computer telephony integration (CTI) or other third-party call control mechanisms (e.g. programmable switches) the connectivity between different users. The originating terminal is able to deal with IP communication and telephony communication. In a first phase

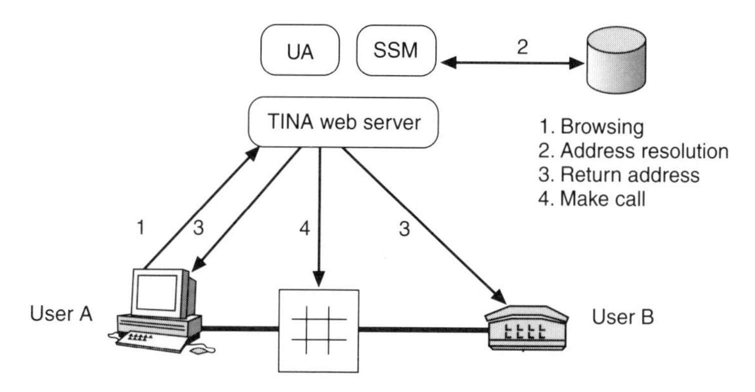

Figure 7.11 A TINA web server controlling the connectivity

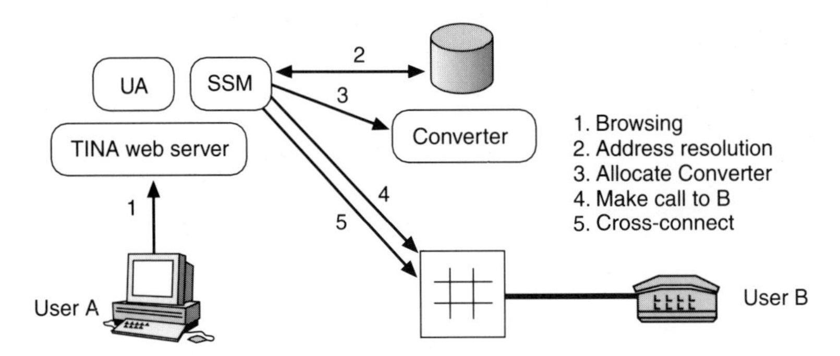

Figure 7.12 A TINA web server converting Internet to telephony

the user A uses a *White Page* application supported by the TINA Web server, then selects a user to call. The TINA Web server uses CTI mechanisms in order to place a call to users A and B. The users are finally connected.

The third option (Figure 7.12) considers a TINA Web server able to use CTI mechanisms to connect to a user and conversion mechanisms for converting voice into IP packets and vice versa (VOIP). In this case the originating terminal is only supporting IP-based communication. When user B is selected, the TINA Web server allocates a conversion unit (a voice gateway), makes a call to user B and forwards the packets from/to the converter.

The fourth option considers a TINA IN server able to offer the IN value (freephone translation, mobility roaming, ...) to a pure Voice Over IP network. Voice Over IP may also mean Multimedia Over IP.

As shown in the previous examples, different degrees of integration are possible between an intelligent network infrastructure and a Web-based telephony server. However, the classes of services that could be supported bypass the telephony services. TINA could be used in order to introduce the concept of session allowing the integration of different transport networks as well as the ability to integrate different sources and sinks of information flows.

7.6.6 Example UMTS

First, some TINA concepts are useful in the short term: for example, the computing platform. Distribution (using CORBA or not) is possible in both TMN and IN parts of UMTS. Prototypes will be available in 1997/98, products could be ready for the UMTS or for the GSM3G, and this evolution is pertinent for scalability reasons.

Second, the TINA architecture is very useful in UMTS and GSM 3G service architecture: it is the medium-term evolution of IN, because the TINA OO service design is the only solution to clear the issues as previously mentioned in Chapter 1. Also TINA provides servers with:

Table 7.1 Possible migration from GSM to UMTS

Product:	GSM	GSM 3G	UMTS
Distribution support:	Existing product	CORBA	TINA DPE
Object model:	Components	Objects (encapsulated components)	TINA objects

- A structured software for very short lifecycle time
- Operators' possible contribution to development, with real multi-providers service co-operation, and with business roles specified interworking at reference points: use of the TINA business model
- Separation of connection, call and service, as this becomes mandatory in such a complex and heterogeneous architecture
- Intelligence co-operation and federation between the different legacy and new network domains, for instance for global mobility achievements, like global roaming features, and global registration with distributed charging.

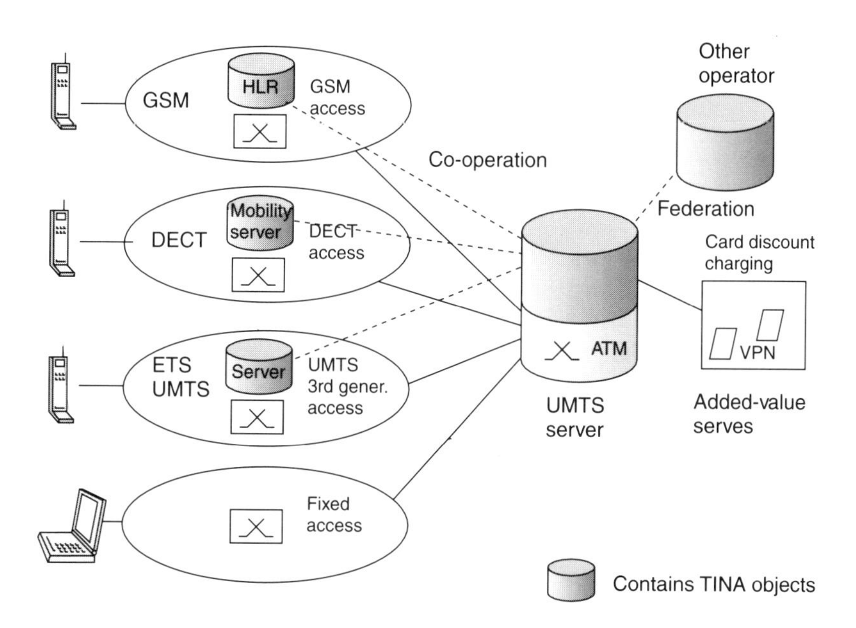

Figure 7.13 UMTS architecture based on TINA

Figure 7.13 illustrates a possible network architecture based on TINA to serve fixed and mobile integration of UMTS.

7.7 TINA IN MANAGEMENT NETWORKS

Network and service management are becoming increasingly important in today's telecommunications. The increasing demand for new or more complex services requires more sophisticated and flexible management systems that can satisfy this demand, guarantee software and network resource integrity, and provide a harmonized and up-to-date view of the network resource state in order to react efficiently and rapidly to any malfunctioning.

The management solutions based on the ITU Telecommunications Management Network M. 3000-series recommendations prove to be effective to manage today's large and complex telecommunications networks comprising equipment from various providers. The international standard is built on the following concepts:

1. Specific reference points between managed and manager systems
2. Modeling of the network element and network information that is required to be shared on both sides of each reference point
3. Computational interfaces at which the managed and manager systems can exchange information and act on one another, specified using an object-oriented language: the Guidelines for the Definition of Managed Objects (GDMO); a single paradigm applies: the manager system acts upon the managed system or receives notification from it, through CMIP (or even SNMP) messages
4. Specific communication platforms based on the "Common Management Interface Protocol" and comprising TMN systems management functions
5. Furthermore, the TMN recommendations include a layering framework in which different management domains have been identified: business management, service management (SM), network management (NM) and network element management (NEM).

However, for the TMN standard to be applicable, more effort needs to be put into information modeling (e.g. access network modeling) and into management platform provisioning. Furthermore, the TMN-based management solutions are not flexible enough and will start to suffer from some drawbacks, of which three are described below:

1. They were associated for too long with a single type of communication protocol, i.e. CMIP or SNMP, whereas the management concepts promoted in TMN do not depend on the underlying communication middleware, and could run, for instance, with the Web-oriented management protocol.
2. They constitute a paradigm shift with respect to the management solutions applied in the computing platform domain or with respect to the service domain,

and therefore they cannot be retained as they are in a framework intended to provide a consistent and global software architecture.

3. They lack interoperability and computing distribution facilities, by which managed objects and manager systems could interwork much better and be distributed more easily on the global management network.

Since TMN-based solutions start to represent a notably important investment for almost all types of transport networks (access networks, SDH transmission, ATM switching) and since they constitute an achievement in comparison with personnel-based monitoring and maintenance solutions, a wise step is obviously to make both TINA and TMN systems cohabit and interwork, and to examine how TINA subsystems can be integrated into TMN networks step by step. No integral substitution of TMN solutions by TINA systems is advocated.

7.7.1 Integration of a GIOP-based kernel network into management networks

There is a general consent that the introduction of a GIOP-based kernel network and middleware in the management networks should be the first step for migrating TMN into TINA. It would not be a revolution, since both are based on similar paradigms [CONC97]: object-orientation, interface descriptions, distributed software, client–server paradigm. Nevertheless, it requires GIOP/CMIP gateways and suitable translation mechanisms as those specified by the Joint Inter-Domain Management (JIDM) Group founded by X/Open and the Network Management Forum (NMF).

7.7.2 Integration of CORBA middleware into management networks

Replacing a CMIP-based platform by an object request broker also requires the introduction of OMG object services dedicated to providing the CMISE features that go beyond the CMIP communication protocol: naming, notification, scoping and filtering. For network management applications built on top of TMN systems management functions, it is not clear whether these functions should be redeveloped using a GIOP-ORB-based communication paradigm or replaced by OMG object services.

7.7.3 Integration of TINA connection management services into management networks

The TINA connection management architecture has been the result of integrating a variety of concepts from the ITU-T open distributed processing reference model X.900, the ITU-T M.30xx recommendations, the network resource information models specified in G.803, etc. Therefore the integration of TINA connection management services into a TMN management network is more a matter of platform interworking, already discussed in the two previous sections.

7.7.4 Integration of TINA end-user services into management networks

Apart from the recurrent concern of having end-user services and management applications interworkable, which was another major requirement identified at the creation of the consortium, the integration of TINA services into management networks raises the question of how to manage services: should end-user services be managed following the TMN concepts and through GDMO interfaces or should service management be more integrated into, and distributed among, service components in a TINA way? The more flexible and generic answer is obviously the latter.

7.8 WRAPPING UP

TINA architecture represents a tool for a unified, overlaid service layer on top of the various existing transport networks such as telephony networks, the Internet, etc. This service layer is decomposed into computing nodes interconnected by a kernel transport network, a common middleware on each computing node and service components or objects contributing to management applications or end-user services, distributed on these various computing nodes. Two major approaches ensure integration of TINA systems and existing networks: adaptation unit and component specialization.

Thus TINA is for the software environment what ATM or IP are for the transport network: a support for multiple services, software services for the former, transport services for the latter. This means that, as for ATM or IP, the decision for deploying TINA or integrating TINA into existing environments will most likely be driven by benefits like cost reduction, economies of scale, re-use, and less by new revenues drawn from new service offerings. This is precisely why TINA will remain a global and collaborative effort: a single actor cannot afford taking the TINA role alone.

Chapter

8 Building TINA

8.1 INTRODUCTION

You should now have a good understanding of the TINA architecture, and of how TINA can be introduced into the legacy telecommunications network. You may have compared it with the existing solutions. If you have been convinced that TINA brings advances in many areas including the openess of networks/platforms, and the reduction of new services release leadtime, then you must be keen to know: How can I apply it, where and when can I apply it? This chapter will help you build this assessment, by displaying the experience gained in the early TINA validation projects, and in the first deployments involving real customers.

Because there has been more than twenty TINA "auxiliary" projects and many company internal test trials, as well as multiple interconnected worldwide demos and TINA services, it is not possible to present an exhaustive list of the results of each project. The few chosen in this chapter do not pretend to be the most significant projects, nor to resume all the others: the worldwide effort spent on such projects among the TINA active companies was above 500 persons per year in 1995 and in 1996. The large-scale applications with more industrial products are now just emerging, and their results are not fully visible. Nevertheless, we expect to have covered enough results to be able to present conclusions on the how, where and when to apply TINA.

Unfortunately, this conclusion is not future-safe: this chapter may be subject to rapid obsolescence. First, the TINA assessment "industrial" trials, for a large part, are still in the learning curve: all the results are not available. Second, the list of TINA products is deemed to vary in a short timeframe. Third, progress in distribution technology and performance has always been very rapid: the *how* to apply will change quickly when new

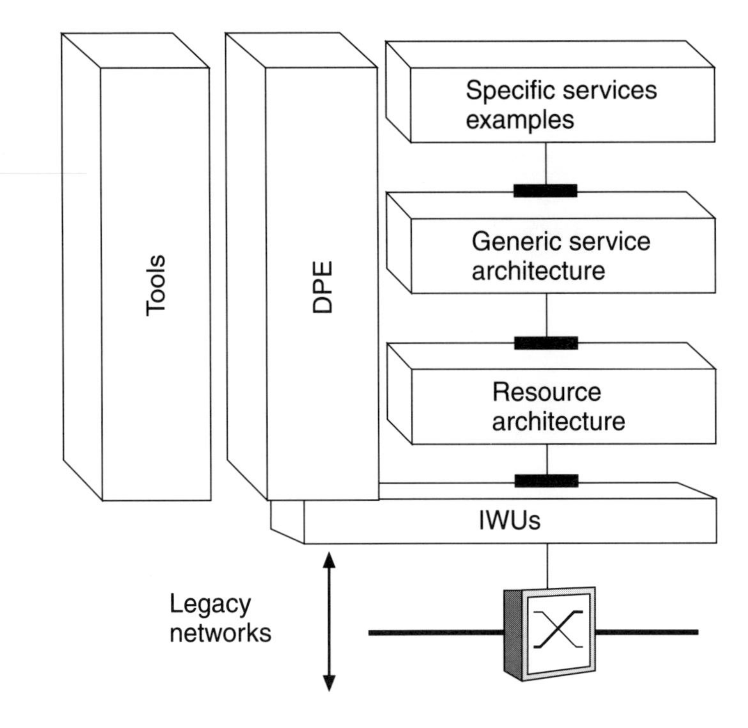

Figure 8.1 TINA scope reference diagram

tools and frameworks are available on the market. The *where* and *when* to apply may also seem too limited within a few years if the performance barriers and the telecommunications specificities have partly vanished.

The chapter is organized into two parts:

- The *how* to apply TINA issue is addressed by section 8.2: we illustrate with a project example the way to create TINA services, and we give some results on the experimentations which have led to the refining, enlarging and specializing of the TINA computing architecture.
- The *where* and *when* issues are addressed by section 8.3.

To ease reading the results from the different implementations, a small diagram locates by a grey area the scope of the experiment, among the general TINA scope areas represented in Figure 8.1. The darkest area shows the major results.

8.2 DEVELOPING TINA APPLICATIONS

8.2.1 Using TINA engineering tools

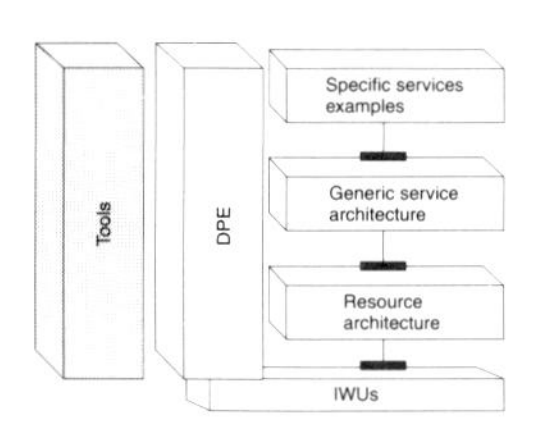

The following summarizes the results of the auxiliary project ACE. In this section we give some hints on how to develop applications according to the TINA model. In Chapter 2, applications were defined in terms of TINA components, the concept of TINA components was defined, and the mapping to the TINA object model was introduced. In this section we describe how to develop TINA components by using software-automated tools that support the TINA object model.

In general, there is not a "standard" methodology to design applications; many approaches can be used: TINA is neither advocating nor rejecting any of them. Adaptations of the most common methodologies and notations (e.g. Booch [BOOC96], OMT [RUMB91], and more recently UML [BOOC97b]) can be considered. Roughly, we can point out different phases in the creation process: requirement gathering, analysis, design, verification, implementation, testing. Below, we give a brief description of the analysis, design and verification phases, showing how a software support tool can help decrease development times and emphasizing the composition of TINA components using the TINA object model.

8.2.1.1 Creation of TINA components

TINA components can be created either "from scratch" or by relying on libraries of existing components. In the "from scratch" case, the approach is typically top-down; in the "re-use" case, it is bottom-up. Of course, hybrid approaches are also possible.

To create a TINA component with a top-down approach typically means starting from some requirements, perform analysis and design and create a computational representation for the component in terms of a computational object specified in ODL (Object Description Language, see Chapter 3). Then the component can be implemented following a variety of CO mappings. Obviously, a pure top-down approach keeps you from the re-use of existing components providing a "part" of the required functionality.

Using a bottom-up approach means relying on *inheritance and aggregation* of existing components, both at the code level (whenever possible) and at the specification level. Whenever aggregation is used, the resulting component is defined in terms of multiple interacting entities: in fact, aggregation means specifying the "composing" entities, the way they interact, and the interfaces that are offered to the outside of the "compound" component. This leads to the definition of a "CO mapping for the component". Note that there can be functionally equivalent structures that use different composing entities.

The best choice is generally to adopt a hybrid approach: a component is created relying on existing components, but adding new parts or using alternative CO mappings to satisfy non-functional requirements.

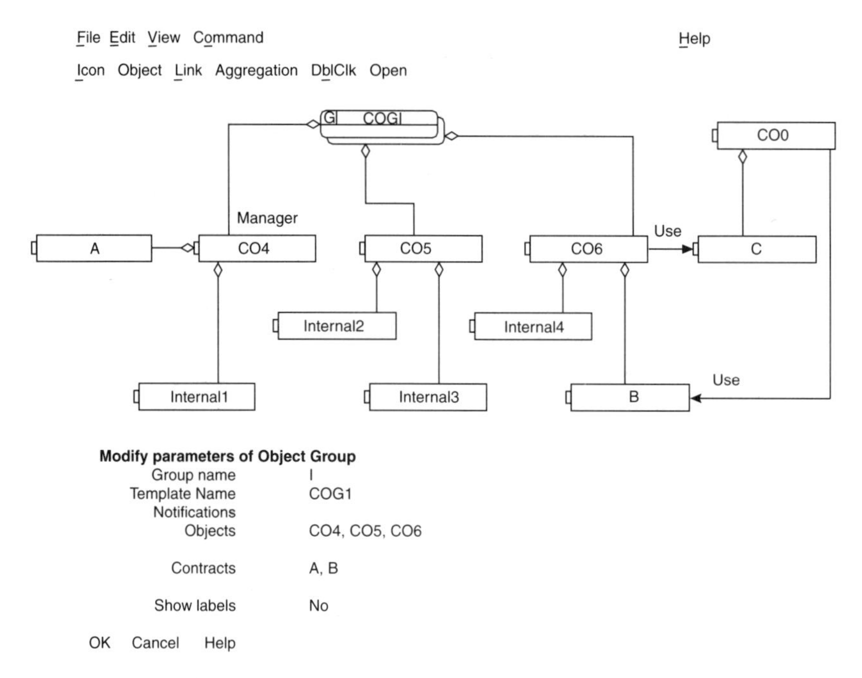

Figure 8.2 Object analysis

8.2.1.2 Automated tools for TINA component development

Software development is a quite complex process, and requires various activities other than design, such as simulation, performance evaluation, etc. In order to develop software according to the TINA model, it is crucial to rely on an automated software tool. Prototypes of TINA-oriented application construction environments (ACEs) have been produced by TINA-C member companies. This section considers the application construction environment prototype developed by CSELT [ACE].

8.2.1.3 Example of TINA component development

In this section we provide some hints on how to develop TINA components using the ACE tool by illustrating a simple example. We adopt a hybrid bottom-up/top-down approach: we assume that we already have specifications and implementation available for some components but that we must map some of them to a different CO mapping.

Let us take as a simple example the service components in Figure 4.10 of Chapter 4. We assume that ODL specifications for IA and UA of the third-party operator are already available, for example because they are part of a TINA component library. Starting from these functional specifications, we can produce program code and perform simulation, using the TINA object model and semi-formal behavior description.

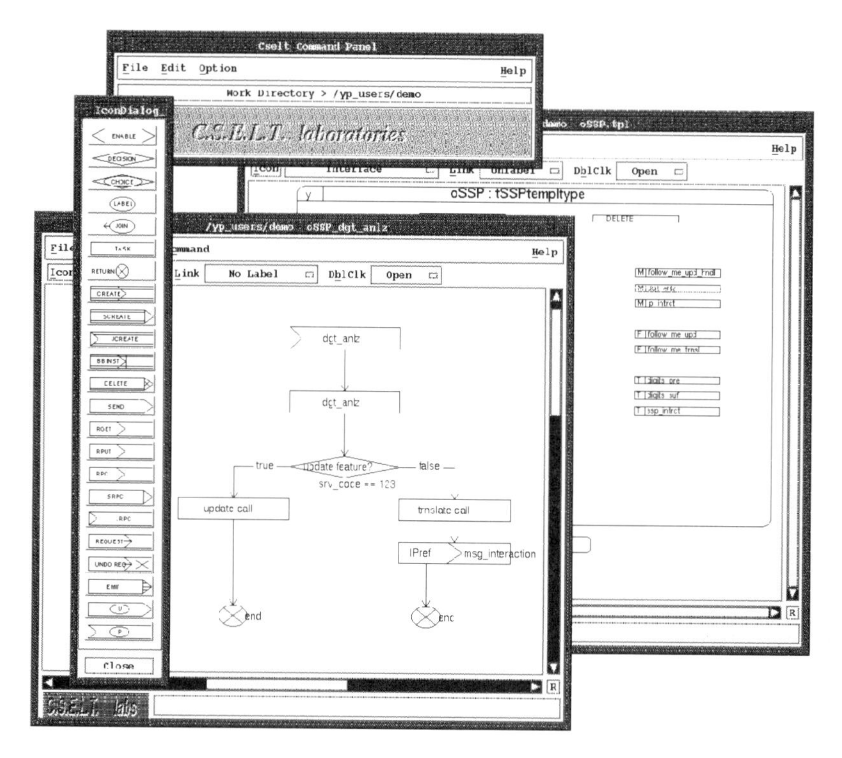

Figure 8.3 Object design

8.2.1.3.1 Analysis

We assume that our requirement analysis leads us to conclude that the component under development can be obtained by combining IA, USM, SSM , SSC, and UA shown in Figure 4.10 in Chapter 4. Using the ACE tool, we draw an initial object model diagram. We adopt the OMT (object modeling technique) notation [RUMB91], properly extended to deal with object groups and interfaces (concepts of the TINA computational model that have no direct equivalent in the standard OMT/UML notation). Each object group, computational object and interface has some associated information (for example, in the case of the object group, what objects are contained, which is the group manager, what interfaces are offered externally as contracts). An example of such a diagram is shown in Figure 8.2.

8.2.1.3.2 Design

After having identified the objects, object groups and interfaces, it is necessary to specify them. This means first, that interface operations are defined in detail, including in/out parameters, etc. In our case, however, we assume that interface specifications are already available (in fact, they are part of the component specifications that we assumed to be

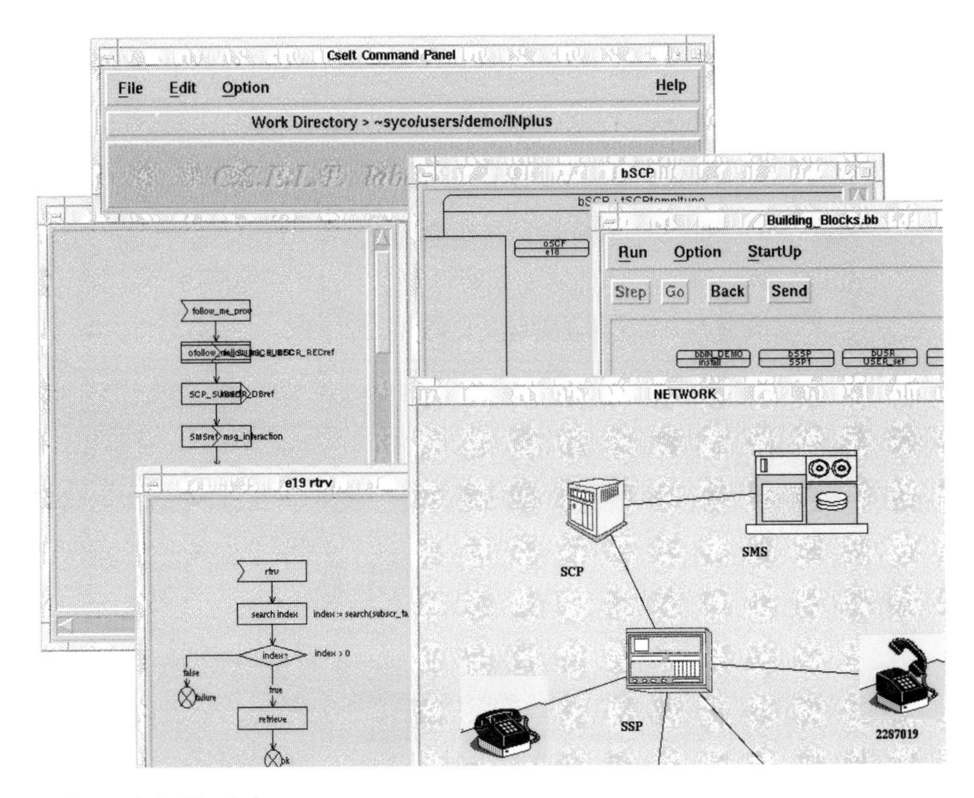

Figure 8.4 Simulation

available), so we can directly import the IDL using our support tool. At this stage, we may want to specialize interfaces, that is, add operations to those already existing, leaving the latter unchanged.

At this point, we are ready to generate TINA ODL specifications for the chosen object structure. Note that these specifications, taken together, are different from the "original" specifications of SC. They include all details of the object group, they specify the interactions between CO4, CO5 and CO6 (i.e. the internal COs in the group) via internal interfaces, etc. Although the language is the same (TINA ODL), the "original" specifications are functional specifications of the component, while these are "structural" specifications of one of its CO mappings.

Now comes the most difficult part of the design phase: the behavior specification. The basis is the ODL behavior description, which consists of plain text. However, a simple text description leaves room for ambiguous interpretation, and it is difficult to be used for verification. Therefore, we need a "semi-formal" graphical language suitable for simulation. First, we map each interface operation to an object method, then we specify methods using the semi-formal language. Figure 8.3 shows an example using an SDL-like notation used with a software tool.

8.2.1.3.3 Verification

Once the service design is completed, it is necessary to verify the correctness with respect to the initial requirements (as well as conformance to standards, if required). A typical way to do this is simulation, which enables us to verify functional aspects; software simulation tools enable this to be performed prior to code generation. Typically, tools like this have to support features such as visualization and tracing of executions. Another important aspect is performance evaluation. This covers a non-functional aspect; software tools support the monitoring of duration of activities, simulating the mapping of objects and messages on specific processing and communication resources, respectively. Figure 8.4 shows a typical simulation screening produced by a simulator used in conjunction with the ACE tool.

8.2.1.3.4 Implementation

The next step is the generation of code targeted to real DPE platforms and programming languages. Software support tools can typically support the generation of C++ code targeted to commercial CORBA platforms, by "translating" the object templates and the semi-formal behavior specifications into CORBA servers.

8.2.2 Building TINA service creation environments

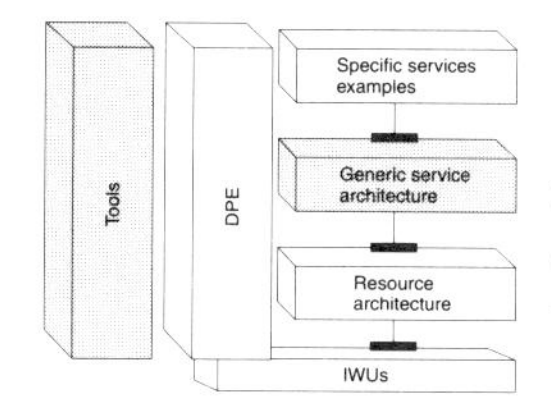

Service creation environments is a well-known domain of intelligent networks, and several IN products are available on the market. Derived from this technology, other service creation environment are now used in the CTI (computer telecommunication interworking) domain of PABX, or in the switching domain. A service creation environment brings the facility to create new services very quickly from a prepared library of service independent building blocks. This is an essential part needed to benefit from a TINA shorter service lifecycle.

Compared to the state of the art of SCE in IN and CTI, service creation of TINA applications is different in many aspects:

- The distribution must be handled with its transparencies
- We must express non-functional criteria like quality of service
- Service interaction should be formalized, and the effectiveness of TINA on this point still needs to be assessed
- The re-usability must be checked at design level as well as at code level
- Along with the CORBA interface specifications, there is still a need to simulate and validate the service call logic.

Meanwhile, the SCE of IN and CTI have not taken full benefit of object-oriented software re-use and ease of development, because the IN and CTI building blocks inherit from a functional design. OO can be used as a language, but does not fit, for instance, with the basic call model in intelligent models. In order to make a proper use of the OO

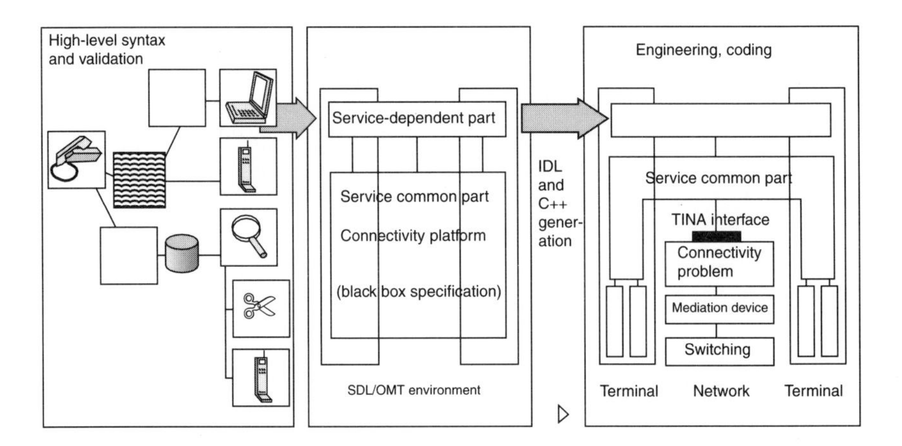

Figure 8.5 Example of TINA service creation environment process

design, one must redesign the call control in an object-oriented view. TINA provides such a model.

It was thus natural that the TINA SCE specifications and prototyping be the subject of research projects. Two auxiliary projects cover this issue: SCREEN and TOSCA. The following summarizes the results from the SCREEN project.

The major issue of TINA SCEs is to produce a seamless toolchain for the creation of TINA service applications, using the object/component-based approach over a distributed infrastructure (Figure 8.5): an SCE complements an ACE with the tooling of the library of components. For instance, part of the TINA components developed in other auxiliary projects like VITAL and DOLMEN are re-used, along with new components added for new applications; all of it is stored in a single component repository. However, the integration of a set of existing and new components is a complex operation, even if a set of common engineering rules and behaviours have been settled.

The pre-existence of functional well-mastered service and call models based on IN or CTI service independent building blocks (SIBs) has led to formal validation chains for the new services design, using SDL simulations: Specification and Description Language (SDL) notation is used.

While using the TINA service architecture, a high-level design based on telecommunication service call modeling still seems to be needed, in order to keep this check of the service logic at a functional level. Living with these two levels of design complicates the analysis, since one has to translate the high-level icon-based service description (*à la* IN CS2 basic call model) into populated TINA object templates, a lower level of the service design.

Because SDL is in essence not object-oriented, there is a need either to redefine simulation and validation of object-oriented call control design or to integrate SDL with object orientation. The integration of SDL use with IDL use is not a notation issue (OSDL will not fit), but a need for active validation (SDL based would fit), keeping the IDL automatic stub generation. The basic design of one TINA object, also called component, can re-use traditional software engineering, as presented in the ACE approach.

In conclusion, up to now, building the TINA service creation environment has been feasible, re-using the previous SDL technology for specification and validation. Nevertheless, it does not provide a simple process with only SDL, or only the object-oriented toolchains. Therefore TINA SCE is a large area for further OO and TINA software engineering research, involving academic work in order to solve the issues detected in the auxiliary projects like the European projects SCREEN and TOSCA.

8.2.3 Refining the computing architecture

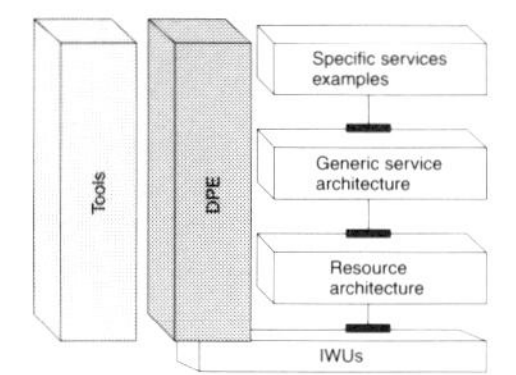

A few TINA projects have had the goal to build an industrial quality distributed processing environment as defined by TINA that would meet a large range of requirements: from the real-time domain up to the management domain. The following resumes the results from the [ReTINA] project.

When using the DPE for the implementation of telecommunication services over various network elements, terminals and management centers, the requirements present a strong heterogeneity in terms of performance, price and feature list. The best way to implement a common computing and distribution architecture in these various environments is to profile the TINA DPE specifications in the different contexts. The following are the results of this profiling.

The minimal real-time DPE called Basic Telecommunications ORB (basic TORB) is meant to serve computational objects in the embedded real-time world, and to be provided on various environments, including a real-time microkernel, Windows NT and UNIX. A richer TORB was obtained by extending the previous one with additional services, including extensive object-based data management, and query facilities, persistency and transactions.

The project has demonstrated that these two profiles were covering most telecommunications development needs. The result is a distributed system product that is scalable and adaptable to different environments, and that has a provision of appropriate performance and quality of service guarantees. To verify this conclusion, the project has developed on top of both TORBs a broadband virtual private networking service that applies the TINA service architecture, the TINA resource and connection management architecture. The two profiles of DPE can be handled in the same toolchain for design, verification and implementation. The project verified this conclusion, using the ACE previously described (Figure 8.6).

Besides the TINA DPE specifications, a few complements appear useful for fine-grain control over real-time resources and scheduling in order to meet the quality of service and real-time guarantee. An example is the logical expression of task management, configuration and persistency management. This was used as an input to the OMG standardization.

Another major need for the implementation and refinement of the computing architecture is the actual performances of real-time ORBs, of the TINA connection manage-

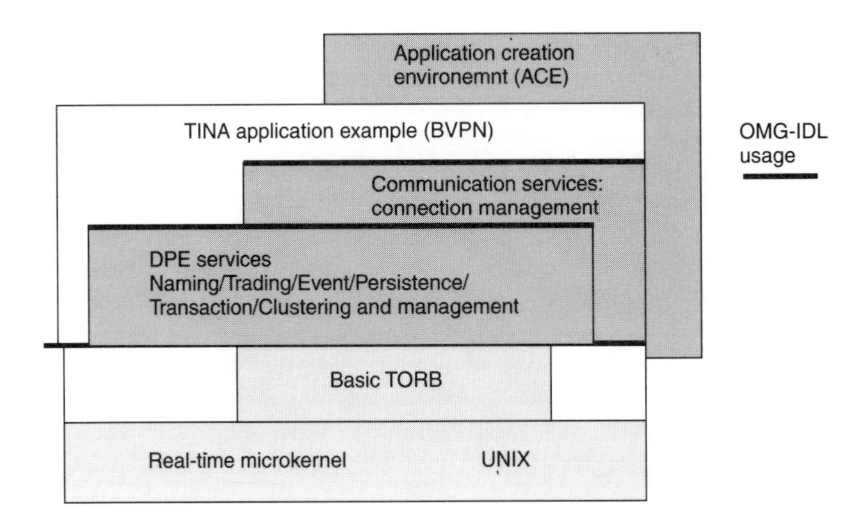

Figure 8.6 DPE profiling = basic versus richer telecom ORB

ment, and of the TINA service management; refinement (retroengineering) to tune the performances to their best is still needed. Among the intermediate results, it was concluded that

- Real-time embedded ORB makes sense as the kernel of future network elements: the performances are acceptable as is the footprint of the basic ORB.
- The TINA architecture developed directly from the specification applying the first generation of development tools requires performance tuning before making an industrial product with it: it is quite natural for a first validation prototype.
- The complements to CORBA required to make proper real-time telecommunication services have been standardized in the OMG.
- Developing a TINA service is considered quick and easy, as compared to traditional methods or to object-oriented methods without architecture models; profiling of the core useful specifications could be done, especially to limit abstraction levels in the flexible binding.

8.3 SERVICE DEVELOPMENTS

8.3.1 A brief history: auxiliary project WWDs and TTTs

The TINA resource architecture and service architecture are a totally new approach for the structuring of telecommunications software. Also the chosen underlying technology for distribution was also new to the information technology domain. There has been much validation activity with most of the TINA consortium members. We can group these into six different sets of trials.

The first set of validations was done by the TINA Consortium Core Team, based in Red Bank, USA, between 1993 and 1997. The second set of trials was carried out inside the TINA-C member companies as own research and development activity and later as private trials or product predevelopment. These are internal, more or less separate activities invisible to the outside world.

It was decided to make TINA feasibility and usefulness visible to the public and also to demonstrate in a third step how easy it is to achieve interoperability. Thus, a few prototypings involving multiple companies in different countries have been conducted. The first occasion to demonstrate such prototypes to the public was the so-called World Wide Demonstrations (WWDs) in Telecom 95 held in Geneva, Switzerland, in October 1995. Preparing these WWDs was the third set of validation projects.

The fourth set was the Consortium's "auxiliary projects", an organized way to get the companies' internal trials advertised, and the reporting on the technical achievements and further TINA refinement needs to the core team. This was the largest part of the TINA effort in 1996–7, looking at the evaluation of 500 man-years invested worldwide. Some of them were partially funded by the European Community through the ACTS (Advanced Communications Technologies and Services) program. Altogether, they contributed significantly to the stabilization of the specification, and to the preparation of the first generation of products.

With nearly full specifications now available, TINA Consortium member companies are developing prototypes that will lead to commercial products. They have been put into trial services in what is called The TINA Trial (TTT) in 1998. This last set of trials also includes the large-scale TINA service implementations advertised as TINA first-generation industrial products.

The sixth set of projects are the newly launched academic co-operation projects, where universities and institutes are used to validate the syntax, models and engineering issues; more generally they will support and provide input to the Consortium's technical workshops for further refinements.

These six sets of trials and developments have contributed to establish the TINA service development's outstanding reputation; they have also contributed to further refinements of the TINA architecture. The following section will give some examples and clues on how to develop services using TINA, based on the results of the WWD, auxiliary projects, TTT and announced products. It is arranged in historical order.

8.3.2 First stage of experimenting: the WWD1 example

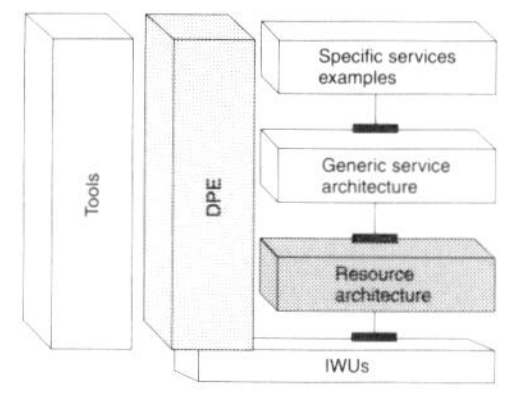

The following resumes the results from the WWD1 project. One can point out in the results of WWD1, the validation of two TINA benefits: multi-operator co-operation in the open resource management architecture and interest in using a trader for configuration management flexibility. The first benefit is one of the major TINA advances. In this domain, the project introduced the original concept of "Telecom Service Agencies".

Also, this large international trial implemented heterogeneous DPE provisioning,

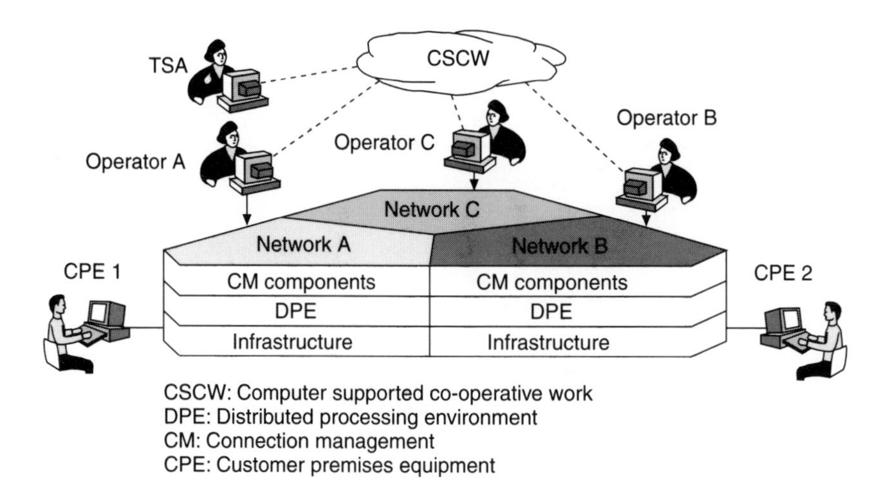

Figure 8.7 Co-operative multidomain network management system

mixing distribution platforms from various vendors. The interoperability of TINA across various hardware and software environments was fully proven in this demo.

8.3.2.1 Service concept of WWD1

A major advantage of the TINA architecture is the ease with which interoperability is achieved in a multidomain service. This was confirmed by developing an experimental multidomain network management service shown in Figure 8.7. In this network model the network management service is provided over multiple networks owned by different network providers using different types of ATM-based network elements. Each network domain uses TINA connection management and DPE.

To achieve prompt service provisioning and network management in a multidomain network, it is essential that users co-operate with network operators. But this is a challenge to users, whose knowledge of networks is quite limited. Thus, users should be assisted by someone called, for example, Telecom Service Agencies (TSA), who can interpret abstract user demand into network management operations and negotiate with network operators. It is the latter that ultimately manage network resources.

Users interact with TSA in a computer-supported co-operative work (CSCW). TSA and network operators also interact in the same environment but additionally have network management systems under their control. Figure 8.8 shows an example of how users, TSA and network operators interact with each other to set up a connection, between CPE1 and CPE2, with capacity requested by one of the users. The denotations in the figure are the same as those described in Chapter 5. A typical connection setup proceeds in the following steps (Figure 8.8):

1. User 1 sends an order to a TSA.

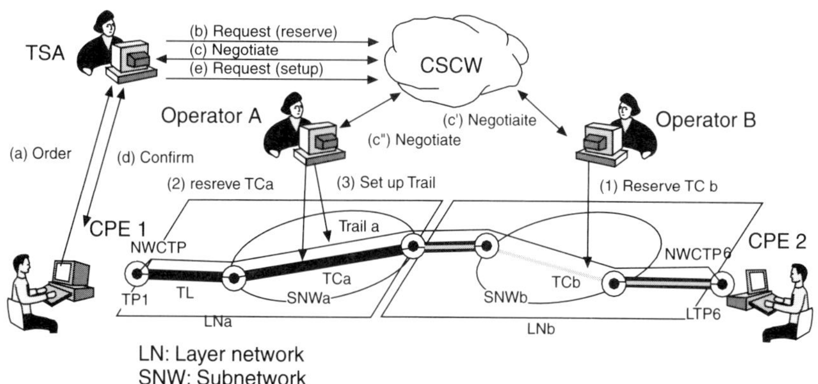

Figure 8.8 Connection setup model

2. The TSA requests connection reservations of network provider operators and negotiates with them to determine the most suitable connection route.
3. The network provider operators reserve connections within their domains.
4. The TSA confirms the feasibility of the order with the user and requests the connection setup of the network provider operators.
5. Finally, a trail is set up over the multidomain network.

8.3.2.2 System Design of WWD1

The system design proceeded as shown in Figure 8.9. Information objects and their relationships are identified for the network resources managed by the network management system. The information model represents an abstract view of network resources. Different operations required on the information objects are identified and computational objects and their interfaces are defined.

The engineering model is developed to define objects in a distributed environment. A trader enables location transparency. Object references are registered in a trader. The co-operative multidomain network management system further requires gateways to support interactions between domains with different DPE runtime environments and interceptors to support interactions between domains with the same DPE runtime environment.

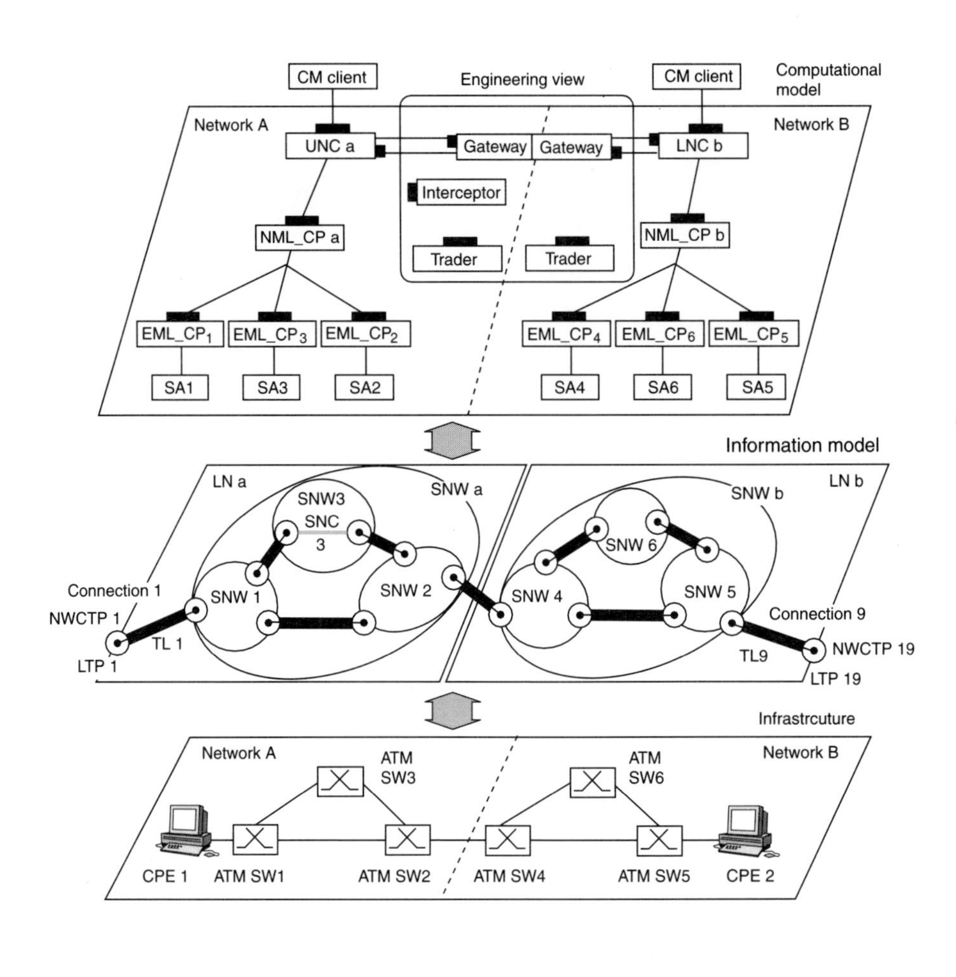

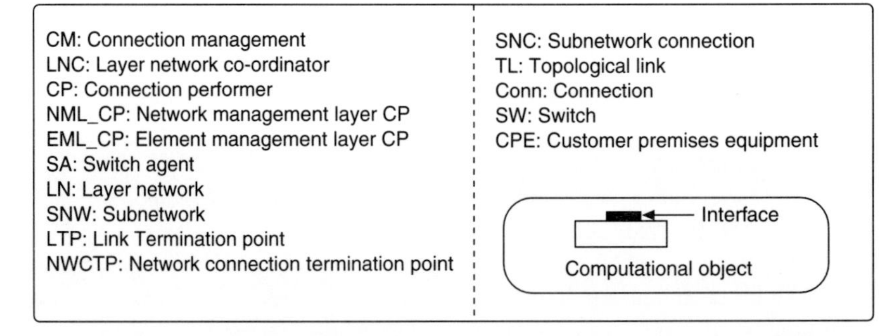

Figure 8.9 System design process

8.3.3 Resource architecture, service architecture and reference points trials

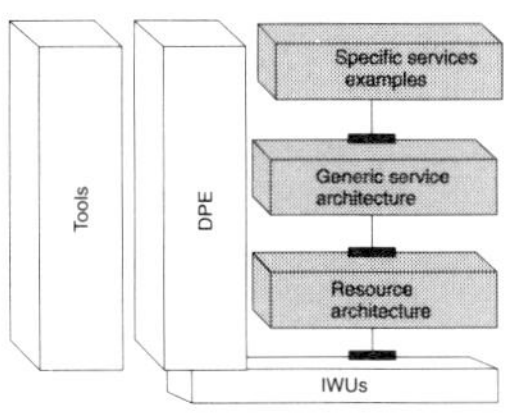

The main goal of this second stage of TINA projects has been to demonstrate and validate in detail the development, deployment, and later management of services based on the TINA service architecture: the service architecture's ease of use, extendibility, scalability, applicability for simple mobile applications, and efficiency in terms of re-usability were checked. A heterogeneous underlying technology was part of most of these deployments, as for the WWDs (Figure 8.10). The following summarizes the results from the [VITAL] project.

There was a deep learning by experience, after implementing several services, including a multimedia multiparty conferencing, a teletraining life application working over the European National Hosts broadband network, and an access mobility service. It is one of the projects providing an extensive experimentation of the whole TINA architecture. The main results were:

- The connection management architecture and the connection graph model are very stable, complete and easy to implement.
- The terminal parts are less stable, for instance there are complementary needs on the TCon reference points for an access graph.
- Re-use of service components from one service to another proved easy, and development time for new service extremely fast (a few weeks for the control part; the management part was about the same).
- Resource configuration was missing (this was done in parallel in the REFORM project).
- The complements to the service architecture for the terminal mobility management were defined and gave conclusions to a 1998 workgroup of the Consortium.
- Implementation was very satisfactory for the first application:

1. Performance of distributed video conferencing was very satisfactory
2. Distributed video-conferencing aspect very well supported by the TINA architecture
   (access – service – communication session concept, session graph concept)
3. Plugging of management applications on TINA "straightforward"

   => VITALv1 has proven the feasibility of the TINA architecture.

8.3.4 Enlarging TINA's scope to network management

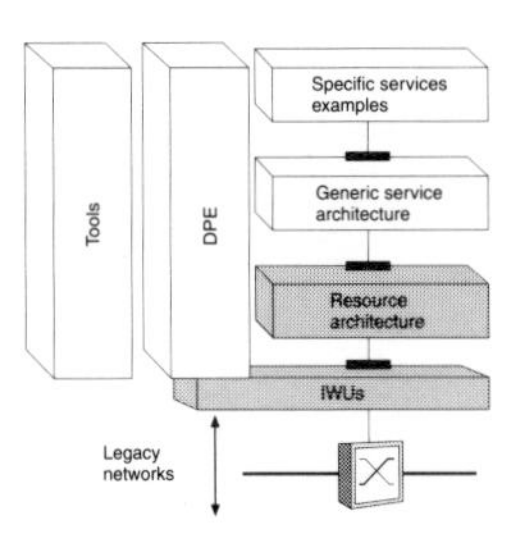

The TINA resource management architecture provides a solution for a large part of the TMN standard configuration management. The rest of the network management standard features (the FCAPS: fault, rest of the configuration like self-healing, accounting, performance and security) can be implemented, using TINA-based object modeling (Figure 8.11). The specifications of such an application of TINA to the network management has not been done in the Core Team, but several projects have worked on this enlargement, like the TINA Trial (TTT by NTT), and REFORM. The following summarizes the results from the [REFORM] project.

Experimentation could bring new results, since the scope of applications is very different from the other TINA trials: it addresses directly the integration of management and control, not only at object modeling but also at run-time co-operation level. The idea is to supervise an ATM network, and to control it in turn, using the TMN features implemented in a TINA fashion.

The TMN features are developed directly over the resource management, and there is no real need for the service architecture. The main results are enlarging specifications for the fault and performance features, for the configuration features for self-healing rep-

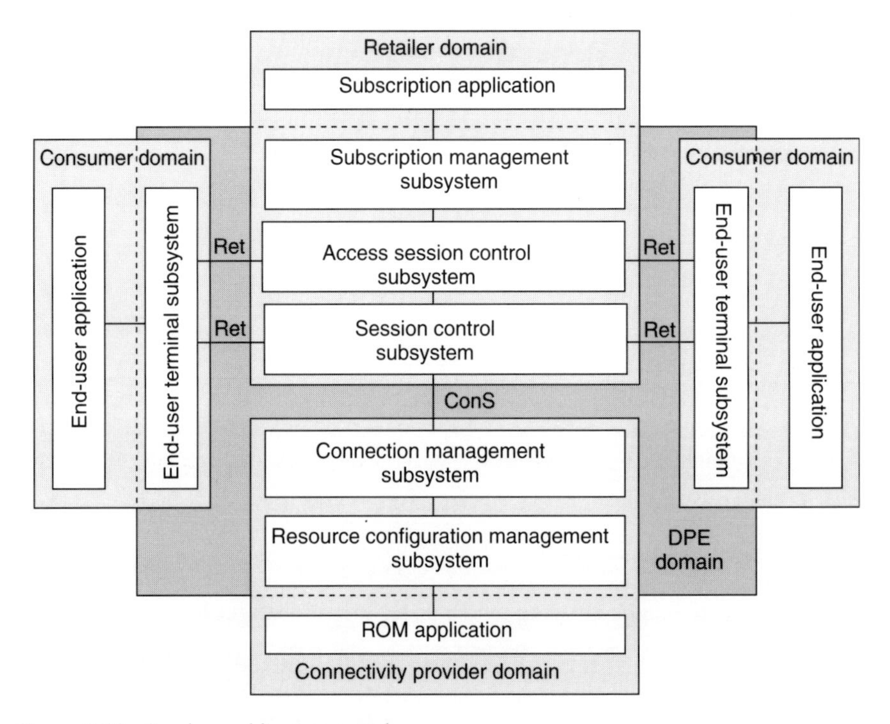

Figure 8.10 Service architecture tested

resentation, and for the relationships between the different blocks: fault to performance, fault to configuration.

Other results have been carried out after investigating the issues related to the migration of TMN legacy networks towards TINA: semantic gaps between the both worlds were analyzed, the question on how much of the TMN paradigm should be translated to TINA was studied. Should event management and FCAPS be strictly translated: statically, with interaction translation?

In conclusion, we can say that applying the TINA computing model to the telecom network management world is a very easy operation, prepared by the object design of the TMN specifications. Going further to the development of TMN features using the TINA architecture does not raise any major issues. Some problems are still visible at the interworking between these new applications and the legacy ones, and are being discussed in the Network Managment Forum.

8.3.5 The TINA Trials (TTT)

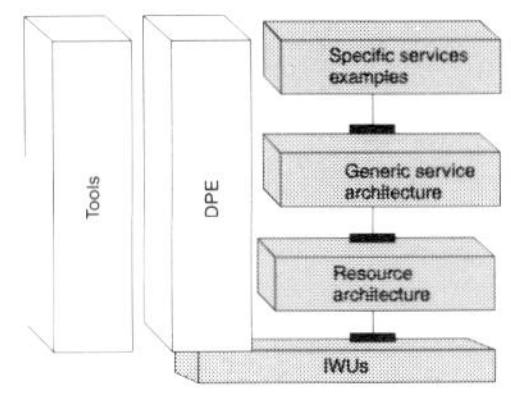

The objective of The TINA Trial is not just a demonstration of TINA feasibility but to be an initial phase for commercial product development. There will be several trials.

The following summarizes the results from the TTT1 project, involving NTT and many other companies. Both service architecture and network resource architecture are involved in the trial. The services include:

- Management services: FCAPS, subscriber management, and customer network management system (CNMS)
- Bearer services: IP over ATM, ATM, non-ATM
- End applications: VoD, on-line shopping

One of the advantages of TINA is that a complex system can be co-operatively developed by many companies without requiring continuous interactions between them. Since software packages can be clearly defined, and their interactions are supported by DPE, they can be separately bought and incorporated into diverse systems. For example, a number of companies may develop certain applications while utilizing connection management and DPE packages developed by other companies as illustrated in Figure 8.12.

8.3.6 Products announced

1997 was a year of announcements, as life TINA deployments were announced:

- For service management purpose with NTT, and with Sprint and Global One
- For ATM CSCW Services with Unisource [SPOT].

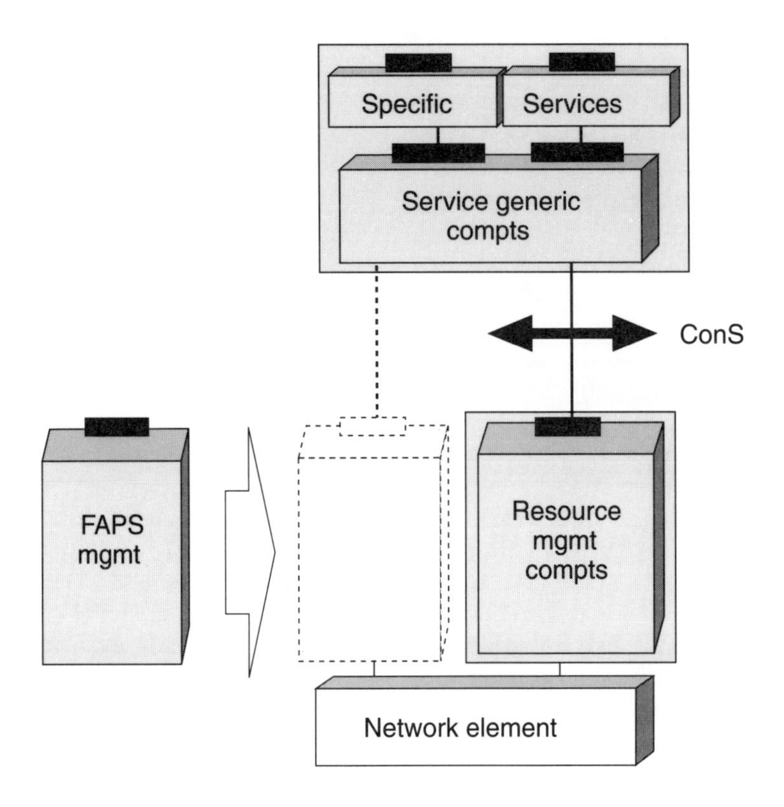

Figure 8.11 TMN FCAPS enlarge the TINA architecture

These larger-scale trials involve long-distance real networks, and life customers that were mainly involved in 1998. They imply a more industrial quality of the TINA components-based products than the previous validation research projects.

8.3.6.1 Example of shared operators and telecom vendor development

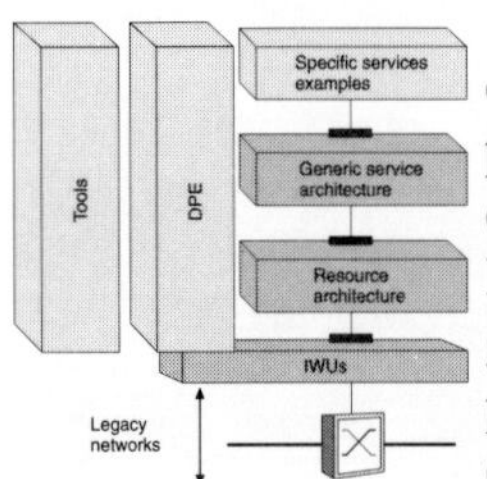

The following gives a few elements of the SPOT trial.

An initial request for information (RFI) entitled "Service Pilot on TINA" was issued at the end of 1995 by the Unisource R&D partners, regarding platforms for a TINA field trial deploying a computer-based co-operative work set of services over ATM. Following the TINA specifications, manufacturers provide heterogeneous network elements and TINA platforms. The operators provide service on top of the open service platform from the telecom vendors. The same terminals have access to services from any vendor and any provider, across an international ATM network.

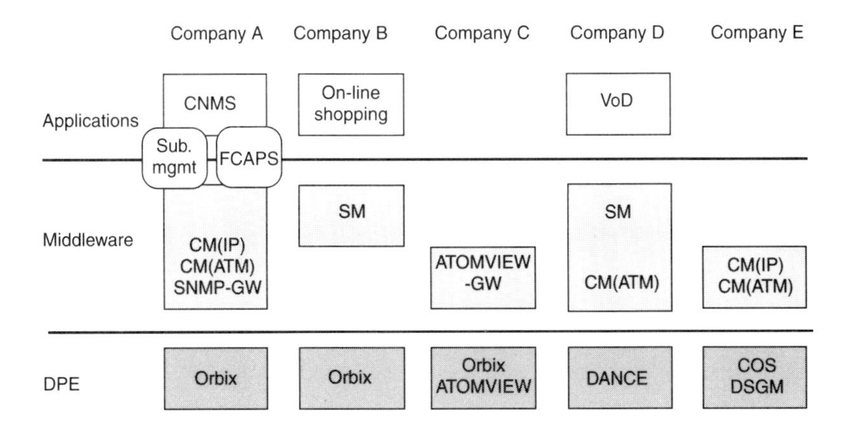

Figure 8.12 Software package development by many companies

Among the results is a clear validation of the openness of the service development platforms, as for the first time part of the service software was developed by operators themselves. The service development leadtime was as short as a few man-weeks. Also, inter-operator co-operation and inter-provider interoperability has been checked, stabilizing the TINA reference points specifications (LNFed, ConS). The cost of industrialization of the TINA previous research prototypes has been in the normal range. Finally, the set of saleable products appears in Figure 8.13.

8.3.6.2 Example of operators product for service management

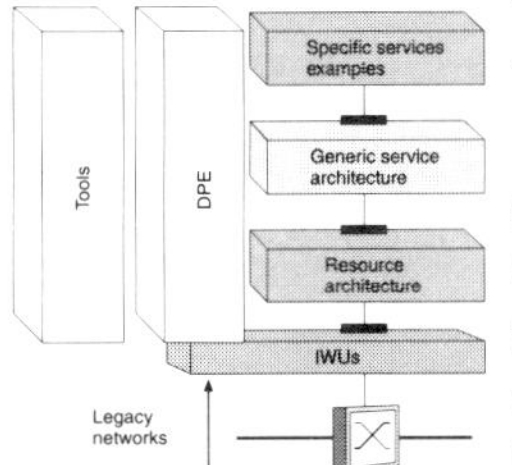

The following gives a few elements of the [GlobalOne Alliance] TINA trial.

The Global One Alliance TINA Trial is a joint effort of the alliance partners Deutsche Telekom, France Telecom, Sprint and Global One to validate TINA-C service and computing architecture in an truly global context. In particular, the work is carried out at locations in Kansas City, Sprints Advanced Technology Labs in Burlingame, at CNET Lannion, CNET Paris, at Deutsche Telekom Berkom, GMD FOKUS Berlin and at Global One's site in Reston, VA. The Trial has been running from September 1997 and continued in its first phase with enhanced features during the year 1998.

The business objectives are to verify TINA-C service component specifications in a multinational environment to proof interoperability and scalability between different implementations of partners involved. The positive outcome and its feedback to enhance the current TINA-C specifications are being used for the implementation of a future Global One Alliance Services Platform. The goal has been to provide a foundation for seamless

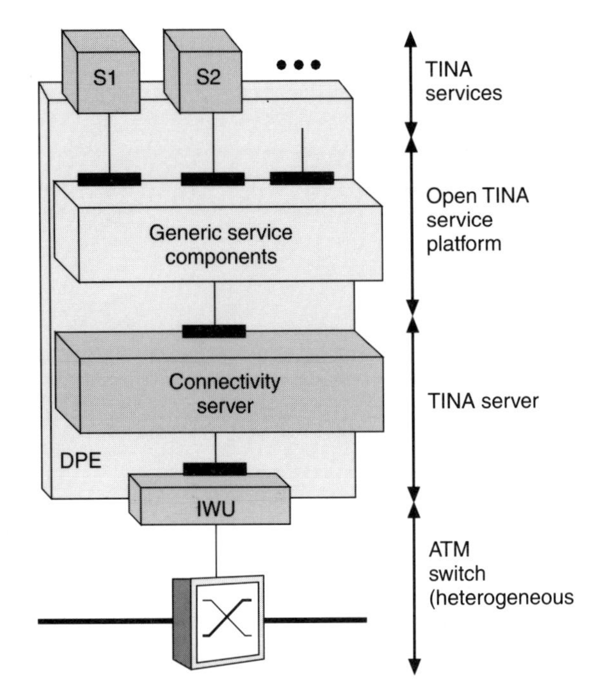

Figure 8.13 Different TINA products to build a complete offer

services in a user-friendly home environment, allowing user and service mobility around the world. The main focus of the trial has been the compliance to customer to retailer (Ret) and retailer to retailer business relationships (RtR), as depicted in Figure 8.14.

The Global One Alliance TINA Trial is based on three types of services:

- Universal access, which allows any user to retrieve his or her accustomed personal service environment from anywhere in the world
- Call completion, which allows a user to be reached anywhere for telephone calls, at whatever endsystem he or she has registered, thus integrating TINA service architecture concepts and POTS and a set of group communication services, such as video conferencing and joint document editing
- In a later stage additional to advanced support for group communication services, the feasibility of service federation will be validated.

The components involved in the access part of the trial are the ones specified in the TINA service architecture, namely the provider agent (PA), the initial agent (IA) and the user agent (UA). It's understood that from the user's point of view there is a single access session only, irrespective of the number of retailers involved. As shown in Figure 8.14, two retailers may be involved. This occurs in situations where the roamed user approaches

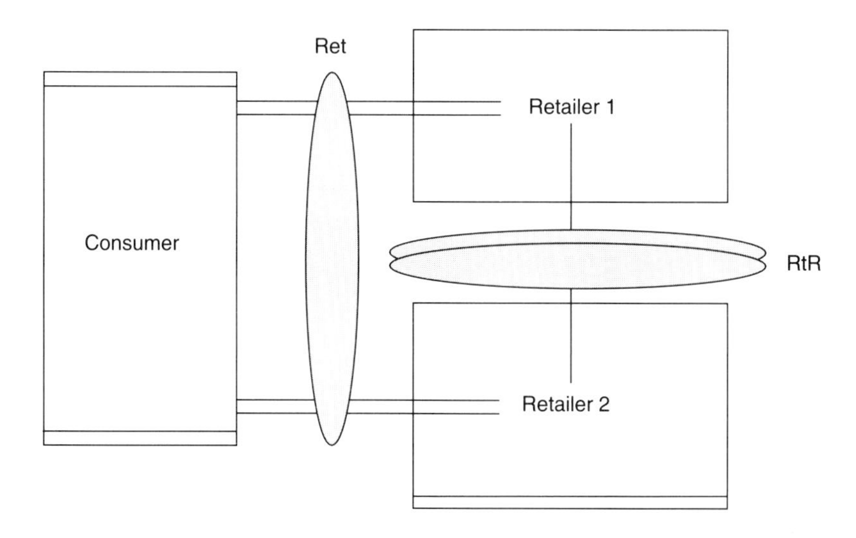

Figure 8.14 Global Alliance TTT, a segment of different administrative domains involved

the first one as a point of presence and visited retailer, whereas the second one is the home retailer. As mentioned above, the purpose of TINA reference point specifications is the definition of generic and open interactions (and interfaces) between stakeholders. Thus the Ret defines the interactions between user and retailer, whereas the RtR reference point defines the interactions between two retailers. When the user establishes the connection to a visited retailer this retailer has to locate the user's home retailer in order to get the user's profile. In the context of the Global Alliance TTT, the retailers are partners and utilize a dedicated alliance kernel transport network. This difference has led to a specialization of TINA specifications, e.g. a different version of the RtR from the one considered in TINA in an "open" environment. Granted that security requirements, for example, are different, reference point specifications can be "lightweighted" or/and enhanced with additional service features for the user. An example would be the listing of invitations to the user since his or her last active access session. Examples of assumptions agreed upon in the trial are:

- Each retailer acts, at least, as a point of presence in the world operated by Global One
- Consumers can access their profile through a local call anywhere in the world operated by Global One
- The retailer acting as a point of presence has to be able to locate the consumer's retailer.
- All retailers (or service providers) involved in the TTT are related together in a Global One Intranet.

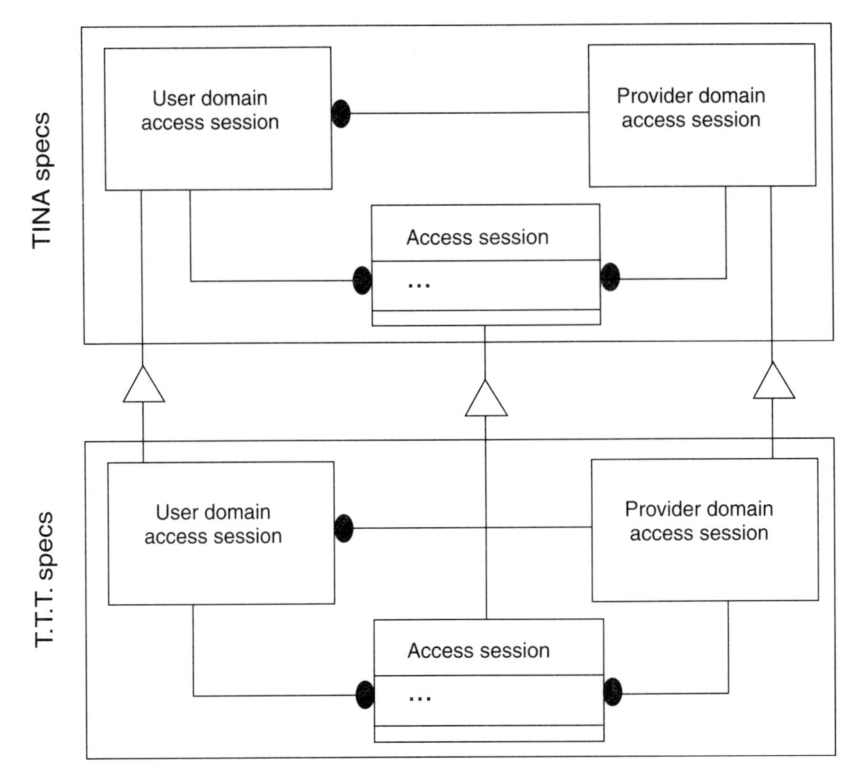

Figure 8.15 Global Alliance's TTT retailer-specific specialization of access session-related components

The specifications of the reference points used in the trials are seen as a specialization but ensure TINA compliance as shown in Figure 8.15.

In a typical scenario of the trial a communication service (e.g. multimedia collaboration) is started through an access session as shown in Figure 8.16. It describes a scenario where in this case a French user visits Germany and uses a laptop with an FT DPE to connect to his or her FT retailer to get access. His or her user agent address is provided by a DT initial agent. The access session invokes a communication service provided by Sprint. The trial runs on an heterogeneous CORBA 2.0 distributed processing environment, consisting of various ORB products from multiple vendors and a dedicated kernel transport network, using ATM and ISDN for international connectivity. Future work is related to service sessions, which run services combining service components of all partners to demonstrate the handling of distributed applications in a global environment.

In summary, the results of the Global One Alliance trial have shown the feasibility of a large-scale experimental implementation of multi-operator services on a multi-retailer TINA platform. The known facts have been demonstrated in the TINA '97 conference in Santiago de Chile, and in the 7th IEEE Intelligent Network Workshop and 5th International Conference on Intelligence in Networks in Bordeaux in 1998.

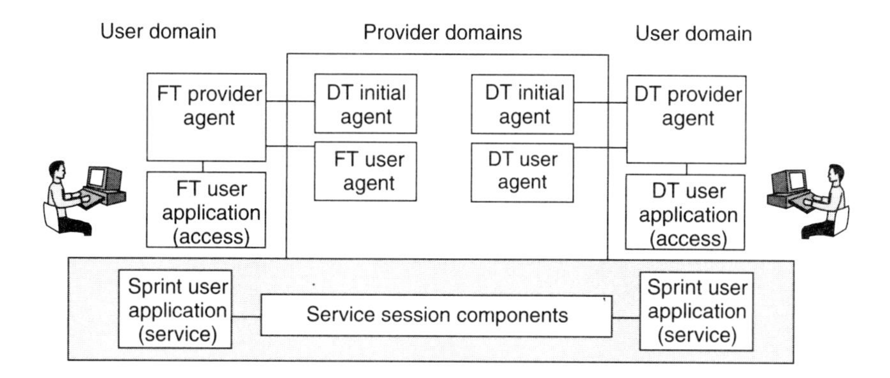

Figure 8.16 A French user in Germany has a conference based on a Sprint service with a German user in Germany

8.4 WRAPPING UP

The implementation of the numerous trials underlined the advantages of the TINA architecture. Because it was possible to clearly define the interfaces of software packages, software development by diverse groups went smoothly. This was particularly notable when, in WWD, SPOT and TTT, companies from different countries and from different business were involved in the development. With some stabilization of the TINA interfaces, it was possible to re-use the components and make them interoperate with each other.
The translation of these validations is that:

- The TINA computing architecture can be applied now, without interoperability problems but with some tooling issues. The few technical issues (naming, etc.) can be solved easily by relying on market solutions.
- The TINA network resource architecture is fairly stable and can also be applied now.
- The service architecture has a rather stable generic core, but in 1997 needed refinement in some areas like mobility, terminals, interworking with legacy precise solutions, and charging, that are being provided progressively in the TINA workgroups: mobility was solved in 1998.

Chapter

9 Conclusions

Chapter 8 has given us some hints of the multiple applications of the TINA architecture and concepts detailed in Chapters 2 to 7. Even if the architecture is a rich (possibly too rich) set of rules in multiple dimensions, prototypes and first developments seem to carry the outstanding conclusion that "all was quick and easy". So what should we believe? Is this magic limited to research centers?

9.1 TINA GEMS AND ROCKS

TINA is a possible business catalyst. Depending on the business perspective and role, various attractions can be found in the computing architecture, in the service architecture, or in the network architecture.

Figure 9.1 summarizes these, which are detailed in the rest of this section. We can further detail the advantages in using TINA for the different stakeholders.

9.1.1 For the service provider and for end-users

- Multi-level customizability of services: TINA introduces a number of levels of customizable services that can be personalized by both the service vendor and the end-user for a number of heterogeneous networks. For example, a service provider can choose to customize a "calling party" service from a library of incoming and outgoing call-screening services, by setting time parameters, charging limitations etc. The end-user can be offered this screening service and customize the "permitted" of the "forbidden" calling parties, depending on the day and on a personal schedule.

- Business opportunity: TINA provides a way to apply services of a network to another network at almost no cost (from GSM to ISDN, from IP to ATM...). It gives seamless service access from different networks, providers and retailers.
- Natural support for mobility and feature interaction: thanks to the generic service object model, TINA permits to apply Roaming in the access part of a call, currently unachievable using Intelligent Networks or ISDN.
- Multivendor multioperator cooperation: TINA reference points provide a frame to define the exchanges between operators; for instance, at the Point of Presence between operators, the TINA network architecture can help legal bodies and multi-operator consorta to define the interactions.
- Software development efficiency: TINA makes producing Telecom software more automated and the development effort more efficient by reusing TINA generic components.
- Product scalability: achieved by the platform distribution using CORBA and the Distributed Processing Environment.
- Software maintenance and robustness: improved by the usage of an Object Oriented framework.
- Internet products enhanced to fulfilling Telecom requirements.
- These goodies can be provided by other solutions, but the availability on the market of standard products will foster the adoption of TINA as the vector of the integration of Information Technologies with a telecom framework. Such an integration needs a timely progress in order to adapt and develop products and technologies to market and in market prices:

  – OO communication: CORBA and DCOM have become a reference similar to UNIX or Windows; IT vendors plan their IN SCP next architecture as distributed using CORBA.

  – OO frameworks: All Service Creation Environments evolve to OO usage.

  – Telecom object libraries: There are emerging standards for plug-in components.

Potential receptors of the TINA goodies for standardization are: OMG, ATM Forum, ITU-T/IN CS3, NMForum, IETF and WWW Consortium.

Figure 9.1 Reasons for migration towards TINA

- Integration of services and management: TINA introduces the possibility of building customized/itemized bills and thus offers operators possibilities for differentiation. It allows operators to give the users an end-to-end view of management of their services through easy (e.g. browser-based) technologies.
- Rapid service creation: The TINA rapid service creation enables both service provider and end-user to benefit from a rapid expansion of services, for instance, on the fixed, mobile and internet convergence market, where there is a large reservoir of possible new service offers.
- Coping with feature interaction: TINA provides a means to cope with the service interaction issue, thus enabling the customer to have one global service environment mixing different services components, like card authentication, virtual pri-

vate numbering and call private filtering, without the side-effects of undesired behavior of the combination.

- Exporting the "network resource API": TINA provides also a high level of integration between network services and customer premises services by means of interactions based on IT protocols and products. In this way a new set of services could be provided to customers that see the network resources and services as programmable entities. In addition, Internet-based GUIs can provide for an easy, but robust way of requesting, controlling and managing services.

9.1.1.1 More specifically for the end-user only

- Integration capability: TINA can integrate in a simple way services provided by multiple service providers over several transport networks. TINA services could, for instance, improve the quality of service offered by the current Internet by orchestrating different network resources in order to ensure and charge the delivery of a service according to the requests of the user.
- Get the best suited service any time: TINA can provide a unique service brokerage, for instance where a broker will be able to sort out the answer to a complex request like finding "which operators provides the end-to-end connectivity at 2Mb/s between Santiago de Chile and London". The service brokerage and the service retailer can be independent of both the terminal type and the network and services operators.
- Openness of TINA-based services: TINA is able to present an integrated and unified service access for several services. This can be useful, for instance, among a variety of services offered by several operators allied in a common consortium.
- Freedom of service composition: As well as using predefined or customized services from service providers, TINA end-users can program together services tailored to their own specific needs at very low effort.

9.1.2 For network and/or service operators

- Reduction of added network and service investment: TINA allows the construction of the operator's own service software, on top of an open interface for any kind of network element. It opens up an entirely new business perspective related to the integration of services and features of heterogeneous networks. Examples are the integration of fixed and mobile networks, as well as Internet and telecom networks. This ability to open service development was a major expectation of intelligent networks; but third-party service providers did not really help for IN, except packaged complete service node solutions involving expensive switching investments.
- Risk sharing: TINA can also be applied to a single network bringing all the potential advantages of an open and programmable architecture made out of components and open APIs. Because TINA is a way for network operators to develop

their own software on standard vendors' equipment, it is a good way to share the development risk between these two parties. Service software deliveries are expected to get a short release lead time, while network element libraries are expected to require longer development cycles. This will allow service providers to focus on service delivery and speed up the service design and provision process, while equipment vendors will improve the network element capabilities guaranteeing backward compatibility and continuity.

- Programmability: TINA provides a number of basic components that can be specialized and enhanced. This feature enables service operators to differentiate through their own software development.
- Straightforward migration: Because TINA organizes a clean cut between network and service applications, it provides a clear migration path for legacy deployed networks. This permits opening a TINA server within a cluster of IN traditional servers; all of them can become distributed around a CORBA software bus.

9.1.3 For equipment suppliers and service companies

- Re-use: TINA provides a componentware-based approach with a generic telecom service modeling. This offers re-use of design and of service software. This is the best way to achieve lightweight development, and to generate margins; in contrast, development costs can be shared on a transparent basis without the need for a single customer to bear the initial burden.
- Get to "state of the art": Because TINA makes use of IT market platforms and tools, it modernizes the telecom software industry where UNIX and object orientation appeared recently on real-time control and services. As it is possible to encapsulate parts of the previous applications and to apply TINA on the upper layer parts only (for instance, services but not connection management) a migration path from existing products is possible.
- Shared development risk: The risk sharing with a third-party object library is also promoted and encouraged; for instance, from IT vendors, a DPE notification server and from telecom start-up companies Signaling System 7 to CORBA interworking units.

9.1.4 For network architects and marketing people

- Clarity of concept: Network architects as well as marketing teams of the operators appreciate having a global service view that covers at the same time service and network management, intelligent network service modeling, network and process addressing flexibility, and fault management and distribution of service functions hidden through transparencies.

9.2 WRAPPING UP

TINA is already proven as a concept in the large number of validation projects which have shown its feasibility in many different contexts. Nevertheless, TINA has not yet reached industrial strength, as some of its components are still young. The same applies to the ORB on which the TINA DPE builds, which has appeared in initial products over the last four years, and is expected to be available in industrial real-time software in two year's time.

As TINA concepts cover a large range of solutions spreading from network architecture and service architecture to Software Architecture, the strategy will not be to have "all-or-nothing" TINA architecture implementation. It is more probable that some of the TINA concepts will be used first in software architecture, then in service architecture, and later possibly in telecommunication network architecture.

Short-term applications of TINA have been shown in intelligent network, network management/service management and mobile location servers. The key aspects of TINA here are distribution and federation, and integrating different network capabilities. Particular attention will be devoted to fixed and mobile network services, and the Internet/Voice Over IP telecom services.

Application of the TINA service architecture as a basis for UMTS mobile systems and for IN long-term architecture is very probable. Long-term application of TINA will be in the evolution of telecom signaling towards IP-based signaling standards with multi-operator reference points.

On the other hand, some operators and manufacturers still doubt the return on investment of the TINA service and network architecture. No one perceives the full TINA network architecture as a short-term implementation guide. It is universally recognized as a target for network evolution, and a brilliant methodological tool for modeling the operator's role. We think that the progressive application of parts of TINA in intelligent networks, in service management and in Voice Over IP will pave the way to the future success of open telecom services and networks.

In order to achieve this success, TINA must seek involvement of vendors and service providers as well as the academic community. The academic world can help to further refine difficult theoretical points such as ATM quality of service handling, and Internet application of TINA. To become part of industrially proved telecom development, TINA needs to grow as an answer to the increasing convergence of information and telecommunication technologies. No doubt it will get to the stage where we use TINA without even knowing it is TINA.

Appendix 1 The TINA Consortium

The first TINA workshop was held in 1990. This was the first occasion to assess in the telecommunications community the need for improving the way services are designed and the common opportunity for tomorrow's services. It was also discovered that similar studies on a software architecture were being conducted in many parts of the world.

These led to the creation in late 1992 of the TINA Consortium (TINA-C) for co-operatively defining a common architecture. The Consortium was created with a five-year life-span, from January 1993 to December 1997. After the end of this (first) phase, a second phase is currently underway.

THE TINA CONSORTIUM 1997–2000

The current form of this Consortium was initiated in December 1997 to continue to further technical refinement, promote the dissemination of the TINA architecture and ensure that it will make business cases. Since the objectives of the Consortium have changed from the creation of the architecture to the creation of business cases from the architecture, the organization of the Consortium has also changed. One of the most important changes is the abandonment of the core team and that the work will be undertaken on the basis of member companies' push. The organization is as follows:

- The Forum is the highest decision-making body of the Consortium and consist of delegates from all member companies. A board of directors takes responsibility for continuing work between Forum sessions.
- Working groups take care of solving the issues (technical, publicity, etc.) raised by the member companies. They are being created to solve specific tasks (and

deleted when the tasks are over) after decisions from the Forum. Such work groups exist at the time of writing, for instance reaching completion of the reference point specifications, the DPE architecture, or the specification of the support for mobility within the architecture

- The Architecture Board (TAB) is in charge of technology aspects, maintaining and guiding consistency between working group results, the outcomes of several other contributing projects and special interest groups, and the overall architecture.
- The home office is in charge of administrative work.

Compared with the organization in 1993–7, the following can be observed. The Consortium Technical Committee and the Consortium Steering Board are terminated and, on the one hand, replaced by the General Forum for decisions on policy and other important matters and, on the other, by the Technical Forum and the Architecture Board for decisions on technical matters. The core team is disbanded and replaced by Technical Working Groups and Special Interest Groups. Auxiliary projects will become contributing projects, focusing on business/applicability aspects.

TINA CONSORTIUM 1993–7

Some 40 telecommunications operators, telecommunications equipment and computer manufacturers joined the Consortium in 1993. A core team consisting of engineers from member companies was brought together in New Jersey, USA, under the assignment to come up with a new architecture integrating both information technology and telecommunication technology. In Telecom 95 in October 1995, TINA-C Worldwide Demonstrations confirmed the feasibility of the TINA principles. As the architecture began to take shape, the people in the home companies have begun to build prototypes to validate the architecture and provide their feedback to the core team.

The TINA Consortium consisted of the following committees, team and office:

- The Consortium Steering Board (CSB) comprises delegates from core member telecommunications operators. It is responsible for defining overall objectives and strategies of the Consortium.
- The Consortium Technical Committee (CTC) comprises delegates from all member companies. It is responsible for all issues for the management of the Consortium to achieve the objectives.
- The Core Team (CT) comprises researchers from the majority of member companies, gathered in New Jersey, USA. It is responsible for defining and validating the architecture.

MEMBERSHIP

The TINA-C membership comprises leading telecommunications operators, telecommunications manufacturers, computer manufacturers and software companies. The membership has steadily grown from the original 33 to about 50 today. We thank them for making this work possible.

TELECOMMUNICATIONS OPERATORS

AT&T, Bellcore, British Telecom, Cable & Wireless, C-DOT, CSELT, Deutsche Telekom AG, Eurescom, ETRI, France Telecom, KDD, KPN, Korea Telecom, MCI, NTT, Portugal Telecom, Sprint, Stentor, Swiss Telecom, Telecom Italia, Telecom Malaysia, Teledanmark, Telefonica, Telenor, Telia, Telstra

TELECOMMUNICATIONS MANUFACTURERS

Alcatel, Ericsson, FUJITSU, GPT, Hitachi, Lucent Technologies, NEC, Nokia, Nortel/BNR, OKI, Siemens

COMPUTER MANUFACTURERS AND SOFTWARE COMPANIES

Broadcom, Digital Equipment Corporation, Hewlett Packard, IBM, IONA Technologies, Samsung, Softwire, SUN Microsystems, Unisys

COLLABORATION WITH OTHER ORGANIZATIONS

The TINA Consortium is not about creating a whole new set of specifications. TINA draws heavily on the results of the latest work in other organizations and standards bodies. The extensions of and new studies made to fill the gaps in existing work are being submitted to these other organizations to achieve global harmony in specifications. In particular, the TINA-C collaborates closely with the ATM Forum, DAVIC, OMG, and ITU-T SG10 and SG11.

Appendix

2 The authors

HENDRIK BERNDT

Dr Hendrik Berndt received his Masters degree in 1971 in Telecommunications and his PhD in 1978 in Fiber Optics from the University of Dresden, Germany. With more than 20 years' experience in the telecommunications industry, he has been involved in a variety of engineering roles including B-ISDN pilot projects on ATM trials and distributed processing technology developments and was head of the Multimedia Application Specifications group in Deutsche Telekom AG. Dr Berndt was the Deutsche Telekom representative to the TINA-C Core Team at Bellcore's New Jersey facility from 1993 to 1996. He was one of the contributing authors and editor of *TINA-C Service Architecture 1995* and liaised with the information and telco industry standardization bodies. After his core-team time he became Executive Director of Advanced Technology for Global One, in Reston, Virginia, the joint venture between Deutsche Telekom, France Telecom and Sprint. He was responsible for migration strategies from IN and TMN to the next generation of distributed object-oriented architectures. Hendrik Berndt is now Chief Technology Officer in the new TINA-Forum.

EMMANUEL DARMOIS

Emmanuel Darmois was born in Paris in 1953. After graduating in Mathematics and Computer Science from Paris University, he became professor in Computer Science at Ecole Nationale des Ponts et Chaussées, a French Grande Ecole (Engineering School) from 1980 to 1991. His research activities were primarily in artificial intelligence and expert systems. He has worked as a consultant and system designer in service creation for the French Minitel. In 1989, he joined Alcatel Telecom in the Research Division as head of a group of engineers working on OO techniques, computing platforms and service creation. He has been the Alcatel representative in TINA-C since the creation of the Consortium, and the TINA Consortium Technical Leader in 1996–7. He is now Director of the Software Division of the Alcatel Research Center.

FABRICE DUPUY

Fabrice Dupuy graduated from the Ecole Polytechnique in 1988 and specialized in telecom engineering at Telecom Paris in 1990. In 1990, he joined the research center of France Telecom (CNET) to work on the benefits of object-orientation and software distribution in telecom (information retrieval) services. In 1992, he contributed to a CNET project on the second generation of intelligent network architecture (TINA, ODP, OODBMS, distributed environments). He joined the TINA-C core team in March 1993, contributing to the TINA-C logical architecture specifications. He was appointed TINA-C associate technical leader in February 1994. In October 1994, he took over the leadership of a team in CNET experimenting with the TINA approach. Since January 1997 he has been heading a R&D department in charge of studying the deployment of TINA-like architecture in FT networks.

MOTOO HOSHI

Mr Motoo Hoshi was born in Yokohama, Japan, in 1944. He received a BS degree in electrical engineering from Yokohama National University in 1968. He joined Nippon Telegraph and Telephone Corporation (NTT) in 1968 and was involved in the development of electronic switching systems. From 1978 to 1986 he carried out research on network digitization planning. From 1985 to 1992, as a Vice-Chairman of CCITT GAS 9, he was also involved in case studies on network digitization in developing countries. From 1987 he has been engaged in research into networking architecture. He was a Core Team member of the TINA Consortium in 1993. He joined the NTT Advanced Technology Corporation in 1996.

YUJI INOUE

Yuji Inoue was born in Fukuoka, Japan, in 1948. He received BE, ME and PhD degrees from Kyushu University, Fukuoka, Japan, in 1971, 1973 and 1986, respectively. He joined NTT (Nippon Telegraph and Telephone Corporation) Laboratories in 1973 and engaged in the development of digital network equipment and the standardization of narrow and broadband ISDN, synchronous digital hierarchy and transport network architecture in ITU-T. While conducting multimedia experiments, he co-initiated the next-generation software architecture called Telecommunication Information Networking Architecture (TINA) at the Consortium of which he has been the Chairperson of its Technical Committee. Dr Inoue is currently the Executive Manager of NTT Multimedia Networks Laboratories. He is a member of IEICE and a senior member of IEEE. He has co-authored several books such as *ISDN* and *Network Architecture*.

MARTINE LAPIERRE

Martine Lapierre is a chief engineer from Ecole Polytechnique and Ecole Nationale Superieure des Telecommunications and has worked for France Telecom for 10 years as a switching engineer and as a Network and Services Planner. She was involved in the early days of the French intelligent network. In 1990 she moved to Alcatel, where she was responsible for broadband marketing and later for the system design of broadband products. She has been in charge of the Information Technology Department of Alcatel Telecom Research for the last five years, where she has formed a large research team on TINA. This team assessed the architecture and prepared TINA-oriented pre-products and strategy. She is now Vice President Product and Marketing in the Switching System Division of Alcatel, and the President of the TINA Forum.

ROBERTO MINERVA

Roberto Minerva works in CSELT's Intelligent Networks department. He graduated *summa cum laude* in 1987 in Computer Science at the University of Bari. At the end of 1987 he joined CSELT, where he was involved in the development of distributed applications for supporting CTI (Computer Telephone Integration). During the period 1989–92 he worked on several European and national sponsored projects for the definition of service architectures (e.g. RACE ROSA, Cassiopeia and R.1044 CSF) and developing software for LAN and ATM interconnection. From 1993 to February 1995 he joined the TINA-C Core Team contributing to service architecture development. In 1994 he was co-ordinator of the Service Refinement group and co-editor of the *Definition of Service Architecture* baseline document. Since 1995, Roberto has been involved in both internal and international activities relating to the application of TINA to actual network systems. From March 1995 to March 1997 he was Project Leader of the EURESCOM P.508 Project, "Evolution, migration path and interworking with TINA", whose goal was to define how and when network operators can benefit from the application of the TINA architecture.

ROBERTO MINETTI

Roberto Minetti has a degree in Electrical Engineering from the Politecnico di Torino (Italy). In 1993 he joined CSELT, the R&D Company of the Telecom Italia Group, working in the Intelligent Network Department. He has been involved in the evolution of software architectures for telecommunications, in the field of both switching systems and service platforms. He took part in several international projects on long-term architectures, within RACE, EURESCOM and TINA-C, participating in the TINA World-Wide Demo 1. From February 1995 to November 1996 he was CSELT's representative on the TINA-C Core Team. In March 1996 he became Champion (co-ordinator) of the Service Stream, the work package responsible for the TINA service architecture. He was editor of the Service Architecture baseline document. At the time of writing, he is Chairman of the Evaluation Group for a TINA-C Request for Refinements/Solutions. Since April 1997, he has been the Project Leader of the "Evolution, Migration Path and Interworking with TINA" project within EURESCOM, the consortium of European network operators.

CESARE MOSSOTTO

Cesare Mossotto was born in Torino, Italy, in 1940. He achieved a full-honours degree in Electronic Engineering at the Polytechnic of Torino in 1964. At present, he is Director General of CSELT (Centro Studi e Laboratori Telecomunicazioni), the research centre of the TELECOM Italia Group. From 1983 to 1988 he was responsible for R&D at Telecom Italia, an Italian public network operator belonging to the STET Group. During his professional career, from 1964 at CSELT and then at Telecom Italia, Mr Mossotto took part in a number of studies and developments concerning traffic theory, digital switching, dedicated data networks and value-added services. Between 1967 and 1984 he was particularly active in standardization bodies, such as CCITT and CEPT, on common-channel signalling systems and ISDN. He is the President of the Steering Board of the Telecommunications Information Networking Architecture (TINA) Consortium, a Member of the ISS Ad Hoc Committee, a member of the board of directors and of scientific and technical committees of several companies and institutions and a member of the Marconi International Fellowship Award Selection Committee. He has also been acting as advisor to the Commission of European Communities (CEC), on several occasions. He is the author of some 60 papers in technical journals and conferences.

HARM MULDER

Harm Mulder received his degree in Electrical Engineering (Micro Electronics) at Twente University of Technology in The Netherlands in 1983. From then until 1993 he worked at KPN Research on network control systems in N-ISDN, B-ISDN and IN. In 1993 he joined PTT TELECOM as account manager for the Network Construction Department. In September 1994 he joined the TINA-C Core Team in Red Bank, NJ, USA, as associate Consortium Technical Leader (aCTL) and later as CTL. His interests lie in network and service control and management, the application of OO methods and distributed platforms, and the business models driving these. He is currently with KPN Telecom Network operations as chief architect on Network and Services for the provisioning of Internet Access Services.

NARAYANAN NATARAJAN

Dr Narayanan Natarajan is a Director in the Internet Architecture Research Laboratory at Bellcore, Red Bank, NJ, USA. He is currently involved in several research projects dealing with broadband network management services and platforms. At Bellcore, he was the chief architect for the Information Networking Architecture (INA) project that developed a distributed processing environment, called INAsoft DPE, and an ATM network management system based on INAsoft DPE. Prior to joining Bellcore, he was at Pennsylvania State University, where he conducted research on distributed real-time systems and high-availability distributed database systems. Prior to that, he was a scientist in the Tata Institute of Fundamental Research, Bombay, where he was involved in several R&D projects in distributed operating systems, multiprocessor systems and concurrent programming. Dr Natarajan has published several papers in the areas of network architecture and distributed systems and has co-authored a book on multiprocessor operating systems.

MAX SEVCIK

After graduating in Electrical Engineering and Computer Science, Max Sevcik became co-founder and principal engineer of Computer Associates, today one of the leading software firms. He was with the Telecommunication Division of Siemens AG in Switzerland and Germany for almost 20 years. With Siemens he held managerial positions in the field of operation and maintenance and in intelligent network systems. His current responsibility is strategic product planning for advanced network services and management for the public switching division of Siemens. With TINA he was part of the project from the very beginning, representing Siemens on the Consortium Technical Committee.

MARTIN YATES

Martin Yates, PhD is an engineer for BT. Since the commencement of the TINA Consortium he has contributed to the design and implementation of internal and collaborative projects using TINA principles and specifications. These have developed service platforms for advanced information and telecommunication services and have employed OMG CORBA distributed object middleware. He led the service architecture work in the Consortium's Core Team during its completion phase from 1996 to 1997. As part of his responsibility he has represented TINA-C output at a variety of meetings and international standardization fora. Dr Yates holds a PhD in the field of optical communications. Prior to TINA, he worked on the analysis and development of object-oriented telecommunications systems and, using this expertise, has contributed to several of BT's strategic systems and planning studies.

Appendix 3 Technical background

This book assumes some technical knowledge in the areas of computer science (mainly object-orientation and CORBA) and telecommunications (mainly intelligent networks and telecommunications management networks). It may happen that some readers may need to be refreshed on some of these topics. A brief overview of these technical areas can be found in this appendix.

This appendix is not intended as an extended tutorial but rather to recall the main characteristics of the technical subjects that are presented. More information can be found in textbooks on these topics, which are given in the Bibliography in the end of this book.

OBJECT-ORIENTED TECHNIQUES

Object-oriented techniques (OOTs) first refer to a new approach (who said "paradigm"? Not us!) for the development of computing systems based on the concept of object. It is important to notice that if object-orientation has been here for more than twenty years, it is only recently that some complex systems based on object-oriented techniques have been delivered (and this is specially true for telecommunication systems). A good text-book on object-oriented techniques has been written by James Rumbaugh *et al.* [RUMB91].

OOTs are now mature and they can be applied at all the stages of the development lifecycle. Developers can, of course, now use well-known OO analysis and design methods and OO programming languages (good point: TINA uses the most mainstream OOAD and OOPL techniques). But they also can base their developments on OO development frameworks, OO databases, OO testing, OO version management, OO project management, etc. (no more excuses for the incompetent OO project managers!).

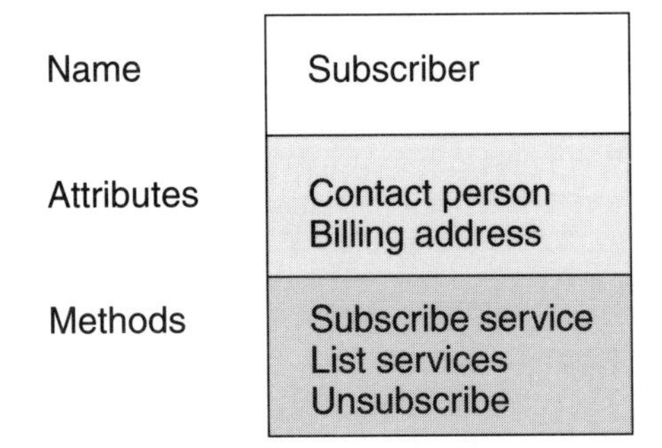

Figure A3.1 Object attributes and methods

A very simple definition of an object is not easy. One of the reasons for the success of objects is that they are analysis/design concepts (the "real world" is a collection of objects: representing it means identifying the objects and their relationships, how an object is made of other objects, etc.) execution supports (pieces of code that have very good characteristics of consistence, self-containment and manageability) and that both points of views are quite close. From design to actual implementation, objects do not suffer those large distortions that are common to the so-called "functional programming". Let us be more specific.

When describing a system, some objects will be defined. Objects have a name or reference (a unique identifier), attributes (specific characteristics of the object than can be changed) and methods (a set of operations that can be performed on the object or that the object can perform). A (very simplified) example of such object is shown in Figure A3.1.

The object described here is a "subscriber", a kind of entity that is dear to the heart of every service retailer since it can subscribe services from them, whether on their own or on behalf of their employing company. The attributes of a "subscriber" are "contact person" (the one that will receive the bill, for instance), and "billing address": both are subject to changes. The methods are quite self-explanatory (this is often the case with object representation!).

Another important aspect of the objects is what they hide. Objects own a set of "private" data, not visible from outside of the object (in particular, not changeable by another object) they have an internal state. This essential feature is called encapsulation.

The "subscriber" object can be seen from different points of view. One can see it as a "template" (a sort of mould) for all subscribers. In this case, one speaks of the class of "subscribers" (one also says type instead of class – though this is sometimes argued in academic discussion but is not within the scope of this section!). One would also like to see a "subscriber" as a particular person (e.g. Jane Doe): Jane Doe is an instance of the class "subscriber". In a computing perspective, an instance will be (can be) an executable object whereas a class is a (design or programming) abstraction.

To design object-oriented systems means to define objects and relationships between objects. This process can be recursive: "smaller" objects can be combined into "larger" objects, which may be used on their turn for even "larger" objects. There are essentially

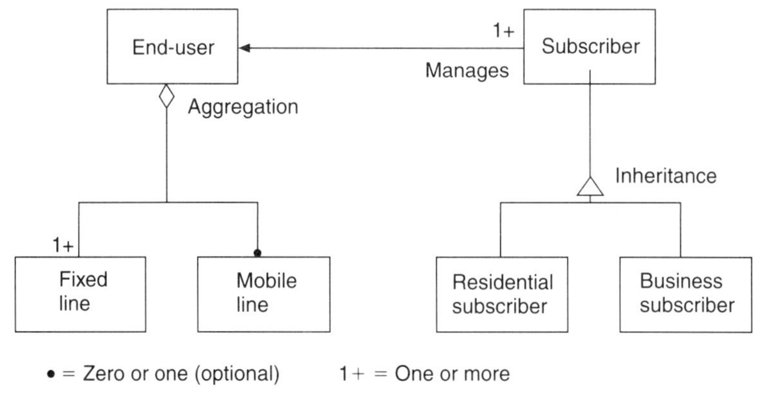

Figure A3.2 Objects: inheritance and aggregation

two ways to do this: specialization and composition. These mechanisms are applied to the classes, not to the instances. They concern the specification level, even though they enable re-use at the code level as well.

Specialization relies on the concept of inheritance: an object class (specialized, or "child" class) is derived from another (*non-specialized* or "parent" class) by refining the attributes and methods. In the above example, the class "subscriber" is specialized into two classes "residential subscriber" and "business subscriber" that have commonalties and also differences. Both classes inherit (the commonalties) from their "mother" class: the attributes and methods of "subscriber" are also defined and available for the inherited classes (unless they are redefined in the child class).

Composition relies on the concept of aggregation: a new object (a compound object) is derived by putting together a set of objects. In Figure A3.2 the "end-user" class is the aggregation of two other classes ("fixed line" and "mobile line') with some constraints on the way the aggregation is done.

Of course, object models are more complex than is described here. TINA has used the mainstream models and methods that are described in [OMT] which is a major reference for OO analysis and design (what TINA is about, even more than OO programming).

CORBA

The rise of object-oriented techniques has one (a real object fanatic would say the only one) major way to design distributed systems in the 1990s and beyond has pushed the information technology industry towards the creation of standards that would allow for the development of interoperable, distributed, object-oriented software applications. This is the objective of the Object Management Group (OMG), a Consortium of more than 700 companies, established in 1989. A good textbook on the basics of CORBA has been written by Mowbray and Zahavi [MOWB95].

The cornerstone of OMG work is the common object request broker architecture (CORBA) which has been designed to allow application objects to discover other objects they need/want to interact with and to over an "object bus". In addition, a vast set of gen-

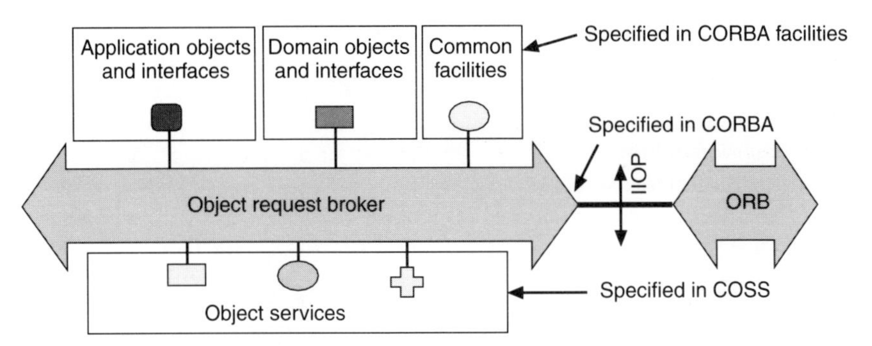

Figure A3.3 The OMA architecture

eric services that support these object interactions has been defined. This set of specifications is grouped in the object management architecture (OMA) presented in Figure A3.3.

OBJECTS AND IDL

CORBA objects are binary components that can be situated anywhere on a network. A client object that requires a service from a server object does not need any knowledge on its actual implementation. What is required (and what is the contractual basis of the binding between the client and the server) is a set of interface specifications. These specifications are written in a neutral language, the Interface Definition Language (IDL), whose role is to describe the interface and structure of the component (attributes, methods, inherited classes, events, exceptions, etc.). Several language bindings (C, C++, Java, SmallTalk, Ada95 and COBOL) allow transparent interoperability (a Java client can interoperate with a COBOL server or the opposite).

CORBA offers static and dynamic services for the client interfaces. In the case of static interfaces, a client object accesses the services via stubs that are the pieces of code on the client side that perform encoding/decoding of the messages exchanged between client and server objects. Market IDL compilers offer an automation of the (highly repetitive) generation of the stubs (one per interface that the client uses from the server). The dynamic invocation interface allows an object to discover methods it can invoke at runtime (offering more flexibility against performance).

ORB: THE OBJECT BUS

The object request broker (ORB) is the center of CORBA (hence CORBA's name!). The ORB facilitates establishing the client/server relationship. A client invokes (statically or dynamically) an interface on a server object (this can be on the same machine or somewhere else in the network). The ORB solves all the issues regarding the location of the server object, the underlying programming language or operating system.

CORBA has been delivered in two major releases. CORBA 1.1 described IDL, the main language bindings and the ORB. This version was meant to offer application portability over different CORBA platforms, but not the possibility of interoperability between objects located on two different implementations of CORBA platforms. CORBA 2.0 has introduced this much-needed interoperability. It relies on two main foundations: the General Inter-ORB Protocol (GIOP), on the one hand, and Repository IDs and Interoperable Object Reference (IOR), on the other.

GIOP specifies the formats (messages and common data) with which interactions between ORBs have to comply. GIOP is meant to work over any connection-oriented transport protocol and can be instantiated for any such protocol. It is no surprise that the most often-used protocol for inter-ORB transport is TCP/IP. The Internet Inter-ORB Protocol (IIOP) is a specialization of GIOP to TCP/IP, and is available from all vendors on the market. Some Environment-Specific Inter-ORB Protocols (ESIOPs) are also used for interoperability over some specific networks: this is used in particular for OSF's DCE.

Interoperability over platforms distributed around a variety of (naming) domains requires that a common naming scheme be proposed for uniquely naming objects and interfaces across platforms and domains. Repository IDs are strings that are system-generated via specific declarations in the IDLs. From an object reference, an ORB creates an IOR (whose format is described in GIOP) whenever the object reference needs to be passed to another ORB.

CORBA SERVICES AND FACILITIES

The basic ORB functionality is enriched by a set of common object services that provide various generic extensions. These services have been specified by the OMG in a series of subsets (COSS1 to COSS5) that provide the specifications for 15 services (at the date of writing) such as: Lifecycle (to manage objects across the ORB), Event (to register/unregister announcements regarding specific events), Naming (to locate other components by their name, making use of standard directory services or naming schemes – e.g. X500 or DCE), Persistency (for object storage), Transaction (offering two-phase commit), Security (to offer authentication, access control lists, etc.), Trader (to allow objects to publish their services and negotiate their jobs with brokers), etc. CORBA facilities are frameworks that can be used directly by application objects such as distributed document composition, mobile agents, business objects, etc.

For more information, one can refer to the standards published by the OMG (Architecture and Specification [OMG-AS], Services [OMG-S], Facilities OMG-F]) or to a number of good books on CORBA such as [MOWB95] (how to design CORBA systems) or [ORFA97] (comprehensive and easy to read, with a very good one-minute introduction to CORBA – what you have just read is a half-a-minute introduction that needs such a good complement).

INTELLIGENT NETWORKS

A set of new services has been requested by customers in the past years that has increased the complexity of the public switching telephony network (PSTN) up to the level where increased complexity would bring down the system. Services were migrated to external services, also to simplify the databases':

- Alternate charging services, e.g. freephone, reversed charging, premium calls, card calls
- Caller guidance: "if you want to order a turkey sandwich, please press 1 now, if you want to order a chicken teriyaki sandwich, . . ."
- Call processing such as screening and rerouting, i.e. the various means offered to subscribers for customizing the way their calls are handled
- Virtual private networks (VPN), i.e. the possibility for a company to have its own private network as a subset of the PSTN.

All these services pose a challenge to a network architecture based on a hierarchy of switches because of the integration of the actual switching function and these features.

The answer to this was proposed in the 1980s and is known as the intelligent network (IN). Basically, the IN architecture is an overlay to the PSTN architecture that provides "intelligence" outside the switches as well as a means of this "intelligence" to control the switches. This allows decoupling of service evolution (mainly software based) from switch evolution (mainly hardware based).

A simple summary of an IN architecture is given in Figure A3.4. Calls are analyzed in the service switching function (SSF) by matching call progress events against triggers. If a trigger fires, the control of the call (together with relevant information) is transferred to an external processor, the service control function (SCF). The SCF will execute a service script, which specifies the action to be taken (e.g. finding the number to be called in case the original number has been forwarded). As a result of the script, the SCF can invoke actions in other IN components, e.g. the intelligent peripheral (IP) to play synthesized voice announcements or the creation of a leg in the SSF. The interaction between IN components is by the Intelligent Network Application Part (INAP) protocol supported by the C7 Transaction CAPabilities protocol stack. The Basic Call State Model (BCSM) is used to describe different stages during the call configuration, separating originating and terminating part of the call.

The ITU-T IN recommendation leaves freedom for the implementor to select a combination of functions in an implementation. The implementations are:
The service control point (SCP):

- The SCF offers the possibility to execute (via a service logic execution environment) the service logic and to mask the underlying computing node
- The service data function (SDF) provides very fast access to subscriber related data.

  The service switching point (SSP):
- The switch performing the PSTN switching function

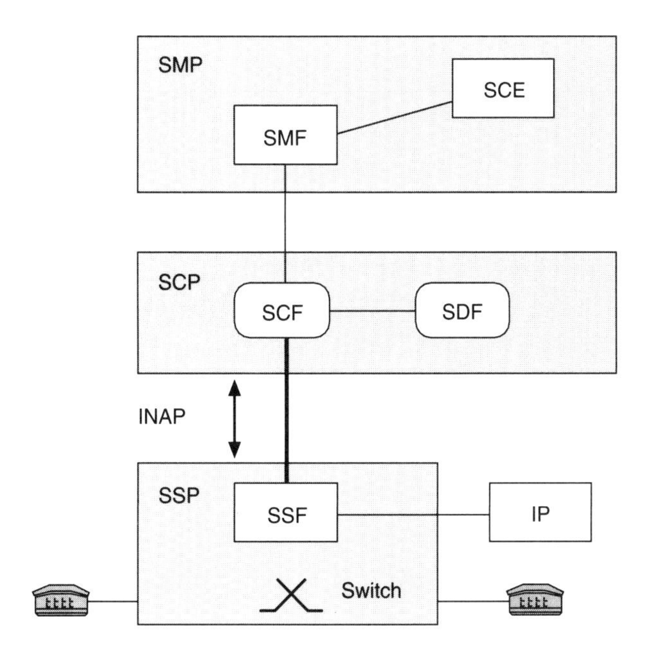

SMP: Service Management Point
SCP: Service Control Point
INAP: IN Application Protocol
SSP: Service Switching Point
SCE: Service Creation Environment
SCF: Service Control Function
SDF: Service Data Function
(based on SS7 signalling+MTP/SCCP/TCAP)
IP: Intelligent Peripheral

Figure A3.4 Intelligent network architecture

- The SSF performing the triggering function and control of the switch.

  The service management point (SMP):

- The service creation environment (SCE), in which a new service logic is created and tested
- The service management function (SMF), which controls the FCAPS management (see the following section on TMN) as well as software distribution management and lifecycle.

This INAP interface is crucial to IN (and is also the most difficult to implement). In the example of Calling Card, the following operations are performed:

- PlayAnnouncement
- CollectInformation (card_number)

- CollectInformation (PIN)
- PlayAnnouncement
- CollectInformation (destination_number)
- ConnectToResource (destination_number)
- CallInformationRequest
- ProvideChargingInformation.

In addition to the SSP and SCP functions on which are imposed very stringent performance and availability constraints, the SMP allows updating of the database of routing information and customer data. The SMP can also be used for SCP software management. The SCE offers the possibility for a subscriber to update information regarding a customer. Typically, an SCE will offer an icon-based interface that will allow the operator to update service data in a transparent way.

The IN architecture can be used for services which go even beyond its initial scope, such as:

- GSM: the architecture for GSM is an example of IN architecture. The functions regarding a mobile subscriber (home location register function that provides the subscription data and the visitor location register that provides the data for the location of the subscriber at a given moment) are provided by servers that can be seen as SCP
- Universal personal telecommunications (UPT): UPT is a set of standardized services designed to offer access to telecommunication services from any network and/or terminal in a uniform way by using a unique personal identifier. The IN architecture is a good candidate for the implementation of UPT.

Intelligent network standards have been initiated by various sources (ITU, ETSI, Bellcore). The ITU standards are provided in increasingly complex capability sets (IN CS-x). IN CS-1 was approved by the ITU in February 1995. IN CS-2 is currently being delayed and IN CS-3 is under discussion. Some TINA concepts have been provided as input for IN CS-3 as part of the IN long term-architecture.

TMN: telecommunications management network

Network operators provide a wide variety of services over different network architectures by using a variety of network elements (NEs). The management of these networks has long been carried out in a completely proprietary way, each operator and vendor having its solution dependent on the network (e.g. PSTN, data networks, mobile networks, etc.) and different NEs provided by different vendors based on different management capabilities and interfaces. With the expansion of the variety of networks and capabilities, network management was turning into a nightmare.

Network operators were working in the 1980s on the definition of principles for

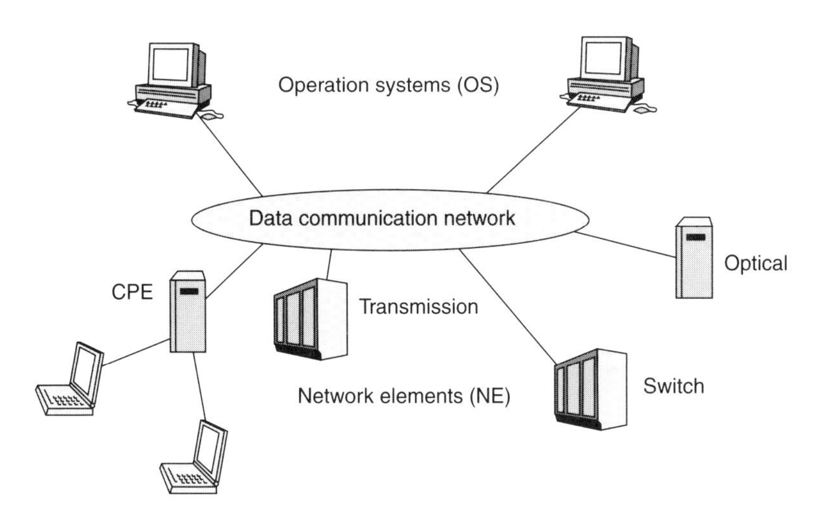

Figure A3.5 TMN Architecture

homogenous network management. These principles are called telecommunication management network (TMN). The underlying model is described in Figure A3.5.

TMN offers a resource-independent abstraction by using an object model of the resource. The operation systems (OS) that manage the network elements do not have to know the details of each different NE. Each NE offers a view of the information that the above layer has to know. The mapping from this information model onto the physical resources within the NE is done by the equipment manufacturer and is not to be known by the OS.

Telecommunications network management is based on a hierarchy of five management layers. Each layer is in a client–server relationship with the next (e.g. the network element management layer is a client of the managed network element layer) which is termed a manager–agent relationship (e.g. the OS is manager for the NE, the NE is an agent for the OS). This corresponds to a hierarchy of operation centers which have the responsibility to operate standard network management functions defined for each layer. The interface between the layers is based on the Common Management Information Protocol (CMIP) and the information exchanged between different layers uses a standardized information model (ITU-T standard M3100).

For each layer, the network management functions are grouped within functional areas termed FCAPS (fault, configuration, alarm, performance, security):

- Fault management deals with the setting and processing of alarms that are emitted by the underlying layer
- Configuration management deals with resource allocation, queries on resource status
- Accounting management deals with the charging information

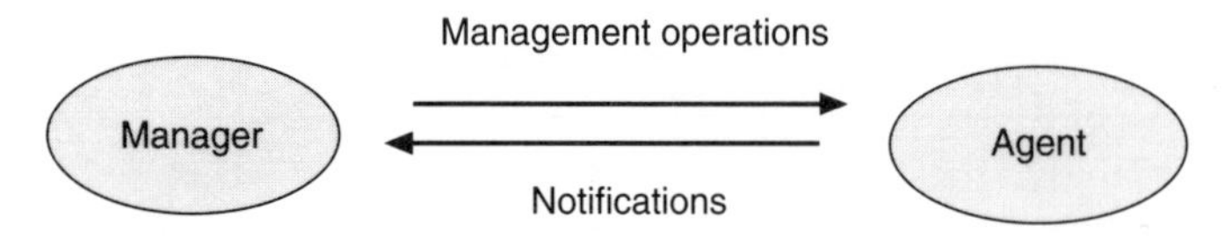

Figure A3.6 Manager–agent relationship

- Performance management deals with setting and logging quality of service information
- Security management deals with defining and enforcing access rights.

The definition of the TMN information model is based on object-orientation. The information model is based on managed object classes (MOCs) and their instances, the managed objects (MOs). The MOCs are defined in a specific language GDMO (Guidelines for the Description of Managed Objects).

MOCs are incorporating a number of packages which are containers for attributes, actions (management operations performed by the manager on the agent) and notifications (reported by the agent to the manager) as shown in Figure A3.6.

The attributes of MOCs are specified with the ASN.1 language. Each MOC instance has to be located uniquely worldwide: this is done via an unambiguous name (the full distinguished name). The MOCs are grouped into management information models (MIMs) that describe the elements of the management information base (MIB).

Appendix

Glossary of acronyms and terms

TINA-SPECIFIC ACRONYMS

Acronym	Definition	Context
3Pty	Third-party reference point	A TINA reference point
AP	Auxiliary projects	Associated TINA-C member company projects
AS	Access session	Service architecture

The objects and relations required to allow a consumer to access services in the retailer domain. An access session does not exist until an access session binding between consumer and retailer domains is established.

Bkr	Broker reference point	A TINA reference point
CC	Connection co-ordinator	Network resource architecture

A computational object in the connection management functional area. It provides clients with the service of interconnection of addressable termination points, multipoint-to-multipoint bidirectional. It hides from clients the concepts of layering and partitioning of transmission networks. The interface specification is based on the connection graph concept.

CO	Computational object	Elements of the computational model

Object in the computational viewpoint. An abstraction that encapsulates data and processing. It provides a set of capabilities that can be used by other objects.

ConS	Connectivity-service reference point	A TINA reference point
CSB	Consortium Steering Board	Management of TINA-C
CSM	Communication session manager	Part of the network resource architecture

A computational object in the connection management functional area. It provides clients with interconnection of computational stream interfaces.

CSMF	Communication session manager factory	
CT	Core Team	A permanent group of resident TINA-C engineers
CTC	Consortium Technical	Technical management of TINA-C Committee (first phase)
DPE	Distributed processing environment	

The abstract infrastructure that provides the execution environment for computationally specified applications, enabling distribution transparencies for distributed applications.

eCO	Engineering Computational Object	Engineering view of a computational object

The engineering representation of a computational object, which encapsulates state/data and processing.

FC	Flow connection	Network resource architecture

A network resource that transports information across a connectivity layer between two or more flow end-points. The characteristic information associated with the different flow end-points of a flow connection may be different.

FCC	Flow connection controller	Network resource architecture
kTN	Kernel transport network	Logically links the DPE nodes

A specific transport network aiming at connecting DPE nodes

LN	Layer network	Network resource architecture

A set of transport functions which support the transfer of information of a characteristic type. Generally, a layer network is closely linked to a specific type of network transmission and/or switching technology, e.g. SDH/SONET VC-4, ATM virtual channel (ATM VC) or ATM virtual path (ATM VP)

NFC	Network flow connection	Network resource architecture
NFEP	Network flow end-point	Element of the NRA Infomodel
NRA	Network resource architecture	
NML-CP	Network management layer – connection performer	Network resource architecture
NWTTP	Network trail termination point	Element of the NRA infomodel

ODL	Object Definition Language	IDL extension to describe TINA objects
PA	Provider agent	Counterpart to UA in the user/provider

A service-independent CO, defined as the consumer domain's end-point of an access session with a retailer.

PSS	Provider service session	
Ret	Retailer reference point	A TINA reference point
RFR/S	Requests for refinement and solutions	TINA-C specifications adoption process
RtR	Retailer-to-retailer reference point	A TINA reference point
SA	Service architecture	
SF	Service factory	Instantiates the service session COs
SFC	Stream flow connection	
SFEP	Stream flow end-point	
SSM	Service session	Service architecture

A specific type of session which models relations or associations between resources and parties in a service. It includes a service-dependent part and a service-independent part. It supports multi-entities negotiation and control, special resource identification and control, and maintenance of the status of the session. It covers connectivity aspects from a high-level point of view only.

SSM	Service session manager	Service architecture
TAB	TINA Architecture Board	
TC	Tandem connection	Network resource architecture
TCon	Terminal connectivity reference point	A TINA reference point
TCSM	Terminal communi-cation session manager	Network resource architecture
TFC	Terminal flow connection	
TINA-C	Telecommunication Information Networking Architecture (Consortium)	
TTT	The TINA Trial	Large-scale TINA-C demonstrations in 1998
UA	User agent	Counterpart of PA in the user/provider relationship

A service-independent CO that represents a consumer in the provider domain. It is the provider domain's end-point of an access session with a consumer.

UAP	User application	
USM	User session manager	Service architecture
USS	User service session	Service architecture

WWD	World Wide Demo	First TINA-C demonstrations in Telecom '95

OTHER ACRONYMS

ALM	NetWare application loadable module	
ANSA		Input to TINA DPE
API	Application programming interface	
ATM	Asynchronous transfer mode	Broadband networking
BCSM	Basic call state model	Intelligent networks
B-ISDN	BroadBand ISDN	
CAMEL	Customer Advanced Mobility Enhanced Logic	Mobile networking
CCF	Connection co-ordinator factory	
CMIP	Common management information protocol	
CMIP	Common Management Interface Protocol	TMN
CMISE	Common Management Interface Service Entity	TMN
COM	(Microsoft) common object model	Object modeling
CORBA	Common Object Request Broker Architecture	OMG architecture
COSS	Common Object Service Specification	OMG architecture
CPE	Customer premises equipment	
CSCW	Computer-supported co-operative work	GroupWare, teleworking services
CSTA	Computer-supported telecom applications	Interface for PABXs
CTI	Computer telephony integration	
CTM	Cordless telephone mobility	
DAVIC	Digital Audio-VIsual Council	Multimedia Services
DCE	Distributed computing environment	
DECT		
DSS	Digital signaling system	
EML	Element management layer	
EML-CP	Element management layer connection performer	
EML-TC	Element management layer topology configurator	
FC	Fault coordinator	
FCAPS	Fault, Configuration, Accounting, Performance Security	TMN
FCC	Flow connection controller	
FLND	Foreign layer network domain	
GIOP	Generic Inter-ORB Protocol	CORBA
GMM	Global multimedia mobility	
GSM	Groupe Special Mobile	The mobile network
HTML	Hypertext Markup Language	
IAP	Internet access provider	
IDL	Interface Description Language	OMG standard

IETF	Internet Engineering Technical Forum	The Internet standardization
IIOP	Internet Inter-Operability Protocol	CORBA
INAP	Intelligent Network Application Protocol	Intelligent networks
IN	Intelligent networks	
IOP	Inter-Orb Protocol	CORBA
IOR	Interoperation Object Reference	CORBA
IP	Internet Protocol	
ISDN	Integrated Services Data Network	The public switched network
ISP	Internet service provider	
ISUP	ISDN Signalling User Part	
IT	Information technology	
ITU-T	International Telecommunications Union	Public telecom standardization organization
JTAPI	Java Telephony Application Programming Interface	
LEX	Local Exchange	
LLND	Local layer network domain	
LNB	Layer network binding	
LNBM	Layer network binding manager	
LNC	Layer network co-ordinator	
LND	Layer network domain	
LNTC	Layer network topology configurator	
LTP	Link termination point	
MAP	Mobile Access Protocol	Mobile networking
MIB	Management information base	
NCA	Network Computing Architecture (trademark of Oracle Corporation)	
NE	Network element	
NEM	Network element management	
NFC	Network flow connection	
NFEP	Network flow end-point	
NLM	NetWare Loadable Module	
NML	Network management layer	
NML-CP	Network management layer connection performer	
NML-TC	Network management layer topology configurator	
NRA	Network resource architecture	
NRIM	Network resource information model	
NTCM	Network topology configuration management	
NWCTP	Network connection termination point	
NWTTP	Network trail termination point	

ODP	Open Distributed Processing	ISO standard
OMA	Object Management Architecture	
OMG	Object management group	
OMT	Object modeling technique	
ORB	Object request broker	The kernel of the CORBA Architecture
OO	Object-orientated (or object orientation)	
OS	Operation system	In TMN
OSF	Operations system function	
PBX	Private Branch Exchange	
PDH	Plesiochronous digital hierarchy	
PPP	Point-to-Point Protocol	
PSTN	Public switched telephone network	
QoS	Quality of service	
RM-ODP	Reference model – open distributed programming	An ISO standard
ROSE	Remote operation service element	
SCE	Service creation environment	Intelligent networks
SCF	Service control function	
SCP	Service control point	Intelligent networks
SDF	Service data function	
SDH	Synchronous digital hierarchy	
SDL	Specification Description Language	
SIB	Service-independent building block	Intelligent networks
SMP	Service management point	Intelligent networks
SMF	Service management function	
SSF	Service switching function	
SSP	Service switching point	Intelligent networks
SS7	Signaling System 7	
TAPI	Telephony Application Programming Interface	Trademark of Microsoft
TCM	Tandem connection manager	
TCP	Transmission Control Protocol	
TCSM	Terminal communication session manager	
TDS	Testing and diagnostic server	
TEX	Transfer Exchange	
TFC	Terminal flow connection	
TLA	Terminal layer adaptor	
TLTP	Topological link termination point	
TM	Trail manager	
TMN	Telecommunications management network	
TSAPI	Telephony Service Application Programming Interface	Trademark of Novell
UDP	User Datagram Protocol	
UML	Unified Method Language	Object-orientation methods

UMTS	Universal Mobile Telecom System	Next-generation mobility standard
UNI	User network interface	
USS	User-to-SCP signaling	
VC	Virtual connection	ATM
VC	Virtual channel	
VCI	Virtual channel identifier	
VoD	Video-on-demand	
VP	Virtual path	ATM
VPI	Virtual path identifier	
VPN	Virtual private network	Intelligent networks
WWW	World Wide Web	Ever heard of this?

Bibliography

TINA-C

The official TINA-C baselines are listed below with the latest approved version (at the date of this writing). TINA-C makes regular updates to these documents, and is likely to introduce new ones (or even to delete outdated ones from this list). More up-to-date information can be found on the TINA-C Web server at:

http://www.tinac.com

1. Architecture

[TC-AS97] TINA-C, "Service Architecture", version 5.0, June 1997.

[TC-AN97] TINA-C, "Network Resource Architecture", version 3.0, February 1997.

[TC-AD94] TINA-C, "Engineering Modeling Concepts (DPE architecture)", version 2.0, December 1994.

[TC-AF96] TINA-C, "TINA Naming Framework", version 0.2, December 1996.

2. Components specifications

[TC-CN94] TINA-C, "Network Resource Components Specification", 1994.

[TC-CI97] TINA-C, "Network Resource Information Model (NRIM)", version 3.0, May 1997.

3. Reference points

[TC-RB97] TINA-C, "Business Model and Reference Points", version 4.0, February 1997.

[TC-RR98] TINA-C, "Retailer (Ret) Reference Point", version 1.0, January 1998.

[TC-RC98] TINA-C, "Connectivity Service (ConS) Reference Point", version 1.0, February 1997.

[TC-RT98] TINA-C, "Terminal Connection (TCon) Reference Point", version 1.0, September 1996.

4. Modeling concepts

[TC-MI96] TINA-C, "Information Modeling Concepts", 1996.

[TC-MC97] TINA-C, "Computational Modeling Concepts", 1997.

[CMCch4]

TINA-C, "ODL Manual", 1996a.

5. Usage

[TC-UG97] TINA-C, "Glossary", version 2.0, January 1997.

OTHER REFERENCES

[AIDA94] S. Aidarous and T. Plevyak (editors), *Telecommunications Network Management into the 21st Century*, IEEE Press, 1994.

[ATMF95] ATM Forum, "CMIP Specification for the M4 Interface (Network Element View)", version 1.0, af-nm-0027-001, September 1995.

[ATMF97] ATM Forum, "M4 Network View CMIP MIB Specification", version 1.0, af-nm-0083-000, January 1997.

[BALB96] G. P. Balboni, F. Bosco, P. G. Bosco, G. Giandonato, A. Limongiello, R. Minerva, S. Montesi, and G. Spinelli, "A TINA-structured Service Gateway", in *Proceedings of TINA'96*, 1996.

[BOOC96] G. Booch, *Object Solutions Managing the Object-Oriented Project*, Addison-Wesley, 1996.

[BOOC97b] G. Booch and J. Rumbaugh, "Unified Modelling Language", Documentation Set, Version 1.0, Rational Software Corporation, 1997.

[BOSC97a] P. G. Bosco, D. Lo Giudice, G. Martini, and C.Moiso, "ACE: An environment for specifying, developing and generating TINA services", in *Proceedings of IFIP/IEEE IM'97*, 1997.

[CONC97] A. Conchon, L. Carré and L. Leboucher, "Does CORBA fit with TMN?" in *Proceedings of ISS'97*, 1997.

[ITUT92a] ITU-T Recommendation G.803, "Architectures of Transport Networks Based on the Synchronous Digital Hierarchy (SDH)", June 1992.

[ITUT92b] ITU-T Recommendation M.3100, "Generic Network Information Model", 1992.

[ITUT92c] ITU-T Recommendation M.3010, Principles for a Telecommunications Management Network, 1992.

[ITUT95] ITU-T Recommendation G.805, "Architectures of Transport Networks", June 1995.

[ITUT96] ITU-T Recommendation G.853-01, "Common Elements of the Information Viewpoint for the Management of a Transport Network", June 1996.

[MOWB95] T. Mowbray and R. Zahavi, *The Essential CORBA: Systems Integration Using Distributed Objects*, Wiley, 1995.

[OMG-A] Object Management Group, "CORBA: Architecture and Specification", 1996.

[OMG-F] Object Management Group, "CORBAfacilities", 1996.

[OMG-S] Object Management Group, "CORBAservices", 1996.

[ORFA-96] R. Orfali, D. Harkey and J. Edwards, *Instant CORBA*, Wiley, 1997.

[REFORM] ACTS project with Alcatel, Algo Systems, HTO, ICS, IONA, NTUA, NTT, Skelton, Telenor, and University College London.

[ReTINA] ACTS project with Alcatel, APM, Broadcom, Chorus systems, CSELT, France Telecom, HP, O2, Siemens, and Telenor.

[RUMB91] J. Rumbaugh, M. Blaha, W. Premerlani, F. Eddy, and W. Lorensen, *Object-Oriented Modelling and Design*, Prentice Hall, 1991.

[SCREEN] ACTS project with Alcatel, France Telecom, INESC, Intracom, Intrasoft, Technical University of Sofia, TeleDanmark, University of Wales Aberystwyth.

[SPOT] Project involving Unisource, Alcatel and Ericsson.

[TINAGate] "A TINA-structured Service Gateway", CSELT.

[VITAL] ACTS project with Alcatel, Belgacom, CSELT, IONA, IT Aveiro, MARI, NTUA, Politecnico di Torino, Portugal Telecom, Telefonica, Universidad de Madrid and University College London.

[WWD1] Wordwide Demo with CSELT, IBM, NTT chapter 8, sec 8.3.2.

Index